BIOZONE Biology Modular Wo...

Ecology

The Biozone Writing Team:
Tracey Greenwood
Lyn Shepherd
Richard Allan
Daniel Butler

Published by:
Biozone International Ltd
109 Cambridge Road, Hamilton 3216, New Zealand

Printed by REPLIKA PRESS PVT LTD

Distribution Offices:

United Kingdom & Europe	**Biozone Learning Media (UK) Ltd**, Scotland Telephone: +44 (131) 557 5060 Fax: +44 (131) 557 5030 Email: sales@biozone.co.uk Website: www.biozone.co.uk
USA, Canada, South America, Africa	**Biozone International Ltd**, New Zealand Telephone: +64 (7) 856 8104 Freefax: 1-800717-8751 (USA-Canada only) Fax: +64 (7) 856 9243 Email: sales@biozone.co.nz Website: www.biozone.co.nz
Asia & Australia	**Biozone Learning Media Australia**, Australia Telephone: +61 (7) 5575 4615 Fax: +61 (7) 5572 0161 Email: sales@biozone.com.au Website: www.biozone.com.au

Front cover photographs:

Autumn leaves, Image ©2005 JupiterImages Corporation www.clipart.com

Clownfish (Amphiprion) ©2006 www.istockphoto.com

Biology Modular Workbook Series

The Biozone *Biology Modular Workbook Series* has been developed to meet the demands of customers with the requirement for a modular resource which can be used in a flexible way. Like Biozone's popular Student Resource and Activity Manuals, these workbooks provide a collection of visually interesting and accessible activities, which cater for students with a wide range of abilities and background. The workbooks are divided into a series of chapters, each comprising an introductory section with detailed learning objectives and useful resources, and a series of write-on activities ranging from paper practicals and data handling exercises, to questions requiring short essay style answers. Material for these workbooks has been drawn from Biozone's popular, widely used manuals, but the workbooks have been structured with greater ease of use and flexibility in mind. During the development of this series, we have taken the opportunity to improve the design and content, while retaining the basic philosophy of a student-friendly resource which spans the gulf between textbook and study guide. With its unique, highly visual presentation, it is possible to engage and challenge students, increase their motivation and empower them to take control of their learning.

Ecology

This title in the *Biology Modular Workbook Series* provides students with a set of comprehensive guidelines and highly visual worksheets through which to explore aspects of ecological theory and practice. *Ecology* is the ideal companion for students of environmental biology, encompassing the basic principles of ecology and field biology, as well as the impact of humans on the natural environment. This workbook comprises six chapters each focussing on one particular area within this broad topic. These areas are explained through a series of activities, usually of one or two pages, each of which explores a specific concept (e.g. food chains or quadrat sampling). Model answers (on CD-ROM) accompany each order free of charge. *Ecology* is a student-centred resource and is part of a larger package, which also includes the **Ecology Presentation Media CD-ROM**. Students completing the activities, in concert with their other classroom and practical work, will consolidate existing knowledge and develop and practise skills that they will use throughout their course. This workbook may be used in the classroom or at home as a supplement to a standard textbook. Some activities are introductory in nature, while others may be used to consolidate and test concepts already covered by other means. Biozone has a commitment to produce a cost-effective, high quality resource, which acts as a student's companion throughout their biology study. Please do not photocopy from this workbook; we cannot afford to provide single copies to schools and continue to develop, update, and improve the material they contain.

Acknowledgements and Photo Credits

Royalty free images, purchased by Biozone International Ltd, are used throughout this manual and have been obtained from the following sources: istockphotos (www.istockphoto.com) • Corel Corporation from various titles in their Professional Photos CD-ROM collection; ©Hemera Technologies Inc, 1997-2001; © 2005 JupiterImages Corporation www.clipart.com; PhotoDisc®, Inc. USA, www.photodisc.com; ©Digital Vision. 3D models created using Poser IV, Curious Labs, 3D landscapes, Bryce 5.5. Biozone's authors also acknowledge the generosity of those who have kindly provided photographs for this edition: • Campus Photography at the University of Waikato for photographs monitoring instruments • Kurchatov Institute for the photograph of Chornobyl • Stephen Moore for his photos of aquatic invertebrates • PASCO for their photographs of probeware • The late Dr. M Soper, for his photograph of the waxeye • Graham Ussher (University of Auckland) for tuatara photos • Dave Ward and Sirtrack Ltd for photos and information on radio-tracking. Coded credits are: **BF**: Brain Finerran (Uni. of Canterbury), **BH**: Brendan Hicks (Uni. of Waikato), **COD**: Colin O'Donnell, **DEQ**: Dept. of Environment QL, **DoC**: Dept. of Conservation (NZ), **EII**: Education Interactive Imaging, **EW**: Environment Waikato, **GU**: Graham Ussher, **JB-BU**: Jason Biggerstaff, Brandeis University, **JDG**: John Green (University of Waikato), **NASA**: National Aeronautics and Space Administration, **NOAA**: National Oceanic & Atmospheric Administration, www.photolib.noaa.gov, **RA**: Richard Allan, **TG**: Tracey Greenwood, **HF**: Halema Flannagan, **KL**: Kevin Lay, Sirtrack Ltd, **UCB**: College of Natural Resources, University of California, Berkeley, **VU**: Victoria University.

Also in this series:

Skills in Biology

Health & Disease

Microbiology & Biotechnology

Evolution

For other titles in this series go to:
www.thebiozone.com/modular.html

Contents

Activity is marked: · to be done; ✓ when completed

How to Use this Workbook

Ecology is designed to provide students with a resource that will make the acquisition of knowledge and skills in this area easier and more enjoyable. An understanding of ecological theory and ecosystem structure and function is important in most biology curricula. Moreover, this subject is of high interest, with many opportunities to combine theory and practical work in the

field. This workbook is suitable for all students of the biological sciences, and will reinforce and extend the ideas developed by teachers. It is **not a textbook**; its aim is to complement the texts written for your particular course. *Ecology* provides the following resources in each chapter. You should refer back to them as you work through each set of worksheets.

Guidance Provided for Each Topic

Learning objectives:

These provide you with a map of the chapter content. Completing the learning objectives relevant to your course will help you to satisfy the knowledge requirements of your syllabus. Your teacher may decide to leave out points or add to this list.

Chapter content:

The upper panel of the header identifies the general content of the chapter. The lower panel provides a brief summary of the chapter content.

Key words:

Key words are displayed in **bold** type in the learning objectives and should be used to create a glossary as you study each topic. From your teacher's descriptions and your own reading, write your own definition for each word.

Note: Only the terms relevant to your selected learning objectives should be used to create your glossary. Free glossary worksheets are also available from our web site.

Use the check boxes to mark objectives to be completed.
Use a **dot** to be done (•).
Use a **tick** when completed (✓).

Periodical articles:

Ideal for those seeking more depth or the latest research on a specific topic. Articles are sorted according to their suitability for student or teacher reference. Visit your school, public, or university library for these articles.

Supplementary texts:

References to supplementary texts suitable for use with this workbook are provided. Chapter references are provided as appropriate. The details of these are provided on page 7, together with other resources information.

Supplementary resources

Biozone's Presentation MEDIA are noted where appropriate.

Internet addresses:

Access our database of links to more than **800** web sites (updated regularly) relevant to the topics covered. Go to Biozone's own web site: **www.thebiozone.com** and link directly to listed sites using the *BioLinks* button.

Activity Pages

The activities and exercises make up most of the content of this workbook. They are designed to reinforce the concepts you have learned about in the topic. Your teacher may use the activity pages to introduce a topic for the first time, or you may use them to revise ideas already covered. They are excellent for use in the classroom, and as homework exercises and revision. In most cases, the activities should not be attempted until you have carried out the necessary background reading from your textbook. As a self-check, model answers for each activity are provided on CD-ROM with each order of workbooks.

Introductory paragraph:
The introductory paragraph sets the 'scene' for the focus of the page and provides important background information. Note any words appearing in **bold**; these are 'key words' which could be included in a glossary of biological terms for the topic.

Easy to understand diagrams:
The main ideas of the topic are represented and explained by clear, informative diagrams.

Tear-out pages:
Each page of the book has a perforation that allows easy removal. Your teacher may ask you to remove activity pages for marking, or so that they can be placed in a ringbinder with other work on the topic.

Write-on format:
You can test your understanding of the main ideas of the topic by answering the questions in the spaces provided. Where indicated, your answers should be concise. Questions requiring explanation or discussion are spaced accordingly. Answer the questions appropriately according to the specific questioning term used (see the facing page).

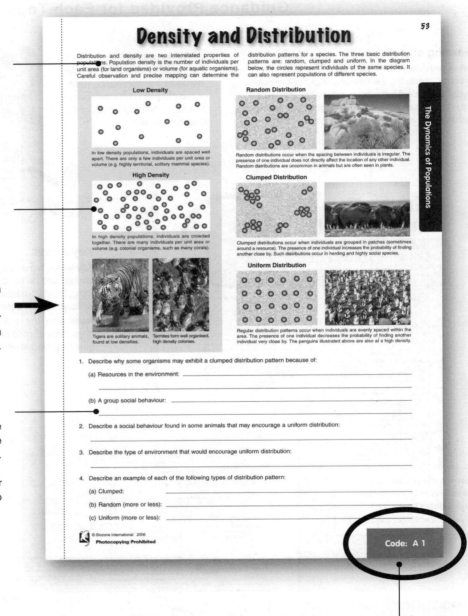

Activity Level
1 = Simple questions not requiring complex reasoning
2 = Some complex reasoning may be required
3 = More challenging, requiring integration of concepts

Type of Activity
D = Includes some data handling and/or interpretation
P = includes a paper practical
R = May require research outside the information on the page, depending on your knowledge base*
A = Includes application of knowledge to solve a problem
E = Extension material

* Material to assist with the activity may be found on other pages of the workbook or in textbooks.

Activity code:
Activities are coded to help you in identifying the type of activities and the skills they require. Most activities require some basic knowledge recall, but will usually build on this to include applying the knowledge to explain observations or predict outcomes. The least difficult questions generally occur early in the activity, with more challenging questions towards the end of the activity.

Explanation of Terms

Questions come in a variety of forms. Whether you are studying for an exam or writing an essay, it is important to understand exactly what the question is asking. A question has two parts to it: one part of the question will provide you with information, the second part of the question will provide you with instructions as to how to answer the question. Following these instructions is most important. Often students in examinations know the material but fail to follow instructions and do not answer the question appropriately. Examiners often use certain key words to introduce questions. Look out for them and be clear as to what they mean. Below is a description of terms commonly used when asking questions in biology.

Commonly used Terms in Biology

The following terms are frequently used when asking questions in examinations and assessments. Students should have a clear understanding of each of the following terms and use this understanding to answer questions appropriately.

Account for: Provide a satisfactory explanation or reason for an observation.

Analyse: Interpret data to reach stated conclusions.

Annotate: Add **brief** notes to a diagram, drawing or graph.

Apply: Use an idea, equation, principle, theory, or law in a new situation.

Appreciate: To understand the meaning or relevance of a particular situation.

Calculate: Find an answer using mathematical methods. Show the working unless instructed not to.

Compare: Give an account of similarities and differences between two or more items, referring to both (or all) of them throughout. Comparisons can be given using a table. Comparisons generally ask for similarities more than differences (see contrast).

Construct: Represent or develop in graphical form.

Contrast: Show differences. Set in opposition.

Deduce: Reach a conclusion from information given.

Define: Give the precise meaning of a word or phrase as concisely as possible.

Derive: Manipulate a mathematical equation to give a new equation or result.

Describe: Give a detailed account, including all the relevant information.

Design: Produce a plan, object, simulation or model.

Determine: Find the only possible answer.

Discuss: Give an account including, where possible, a range of arguments, assessments of the relative importance of various factors, or comparison of alternative hypotheses.

Distinguish: Give the difference(s) between two or more different items.

Draw: Represent by means of pencil lines. Add labels unless told not to do so.

Estimate: Find an approximate value for an unknown quantity, based on the information provided and application of scientific knowledge.

Evaluate: Assess the implications and limitations.

Explain: Give a clear account including causes, reasons, or mechanisms.

Identify: Find an answer from a number of possibilities.

Illustrate: Give concrete examples. Explain clearly by using comparisons or examples.

Interpret: Comment upon, give examples, describe relationships. Describe, then evaluate.

List: Give a sequence of names or other brief answers with no elaboration. Each one should be clearly distinguishable from the others.

Measure: Find a value for a quantity.

Outline: Give a brief account or summary. Include essential information only.

Predict: Give an expected result.

Solve: Obtain an answer using algebraic and/or numerical methods.

State: Give a specific name, value, or other answer. No supporting argument or calculation is necessary.

Suggest: Propose a hypothesis or other possible explanation.

Summarise: Give a brief, condensed account. Include conclusions and avoid unnecessary details.

In Conclusion

Students should familiarise themselves with this list of terms and, where necessary throughout the course, they should refer back to them when answering questions. The list of terms mentioned above is not exhaustive and students should compare this list with past examination papers / essays etc. and add any new terms (and their meaning) to the list above. The aim is to become familiar with interpreting the question and answering it appropriately.

Using the Internet

The internet is a vast global network of computers connected by a system that allows information to be passed through telephone connections. When people talk about the internet they usually mean the **World Wide Web** (WWW). The WWW is a service that has made the internet so simple to use that virtually anyone can find their way around, exchange messages, search libraries and perform all manner of tasks. The internet is a powerful resource for locating information. Listed below are two journal articles worth reading. They contain useful information on what the internet is, how to get started, examples of useful web sites, and how to search the internet.

- **Click Here: Biology on the Internet** Biol. Sci. Rev., 10(2) November 1997, pp. 26-29.
- **An A-level biologists guide to The World Wide Web** Biol. Sci. Rev., 10(4) March 1998, pp. 26-29.

Using the Biozone Website: www.thebiozone.com

The **Back** and **Forward** buttons allow you to navigate between pages displayed on a www site

The current **internet address (URL)** for the web site is displayed here. You can type in a new address directly into this space.

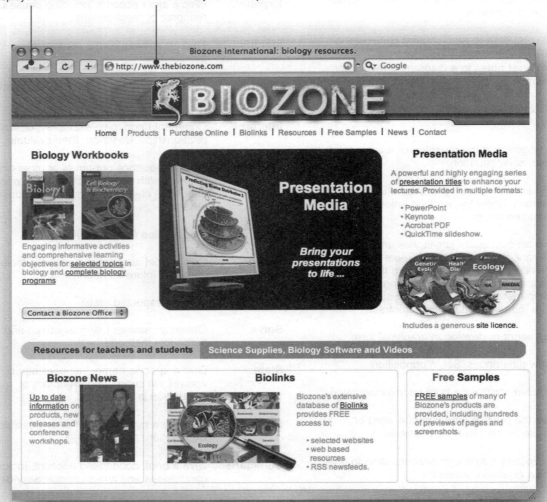

Biozone International: biology resources.

http://www.thebiozone.com — Q▾ Google

BIOZONE

Home | Products | Purchase Online | Biolinks | Resources | Free Samples | News | Contact

Biology Workbooks

Engaging informative activities and comprehensive learning objectives for <u>selected topics</u> in biology and <u>complete biology programs</u>

Contact a Biozone Office ▾

Presentation Media

Bring your presentations to life ...

Presentation Media

A powerful and highly engaging series of <u>presentation titles</u> to enhance your lectures. Provided in multiple formats:

- PowerPoint
- Keynote
- Acrobat PDF
- QuickTime slideshow.

Includes a generous **site licence**.

Resources for teachers and students — Science Supplies, Biology Software and Videos

Biozone News

<u>Up to date information</u> on products, new releases and conference workshops.

Biolinks

Biozone's extensive database of <u>Biolinks</u> provides FREE access to:

- selected websites
- web based resources
- RSS newsfeeds.

Free Samples

<u>FREE samples</u> of many of Biozone's products are provided, including hundreds of previews of pages and screenshots.

Searching the Net

The WWW addresses listed throughout this workbook have been selected for their relevance to the topic in which they are listed. We believe they are good sites. Don't just rely on the sites that we have listed. Use the powerful 'search engines', which can scan the millions of sites for useful information. Here are some good ones to try:

Alta Vista:	**www.altavista.com**
Ask Jeeves:	**www.ask.com**
Excite:	**www.excite.com/search**
Google:	**www.google.com**
Go.com:	**www.go.com**
Lycos:	**www.lycos.com**
Metacrawler:	**www.metacrawler.com**
Yahoo:	**www.yahoo.com**

Biozone International provides a service on its web site that links to all internet sites listed in this workbook. Our web site also provides regular updates with new sites listed as they come to our notice and defunct sites deleted. Our BIO LINKS page, shown below, will take you to a database of regularly updated links to more than 800 other quality biology web sites.

The Resource Hub, accessed via the homepage or resources, provides links to the supporting resources referenced in the workbook. These resources include comprehensive and supplementary texts, biology dictionaries, computer software, videos, and science supplies. These can be used to enhance your learning experience.

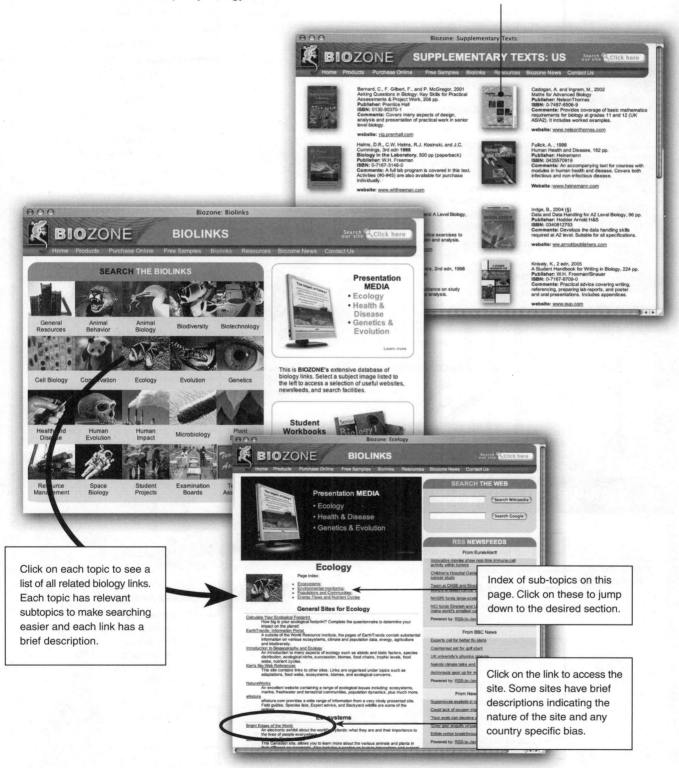

Click on each topic to see a list of all related biology links. Each topic has relevant subtopics to make searching easier and each link has a brief description.

Index of sub-topics on this page. Click on these to jump down to the desired section.

Click on the link to access the site. Some sites have brief descriptions indicating the nature of the site and any country specific bias.

Concept Map for Ecology

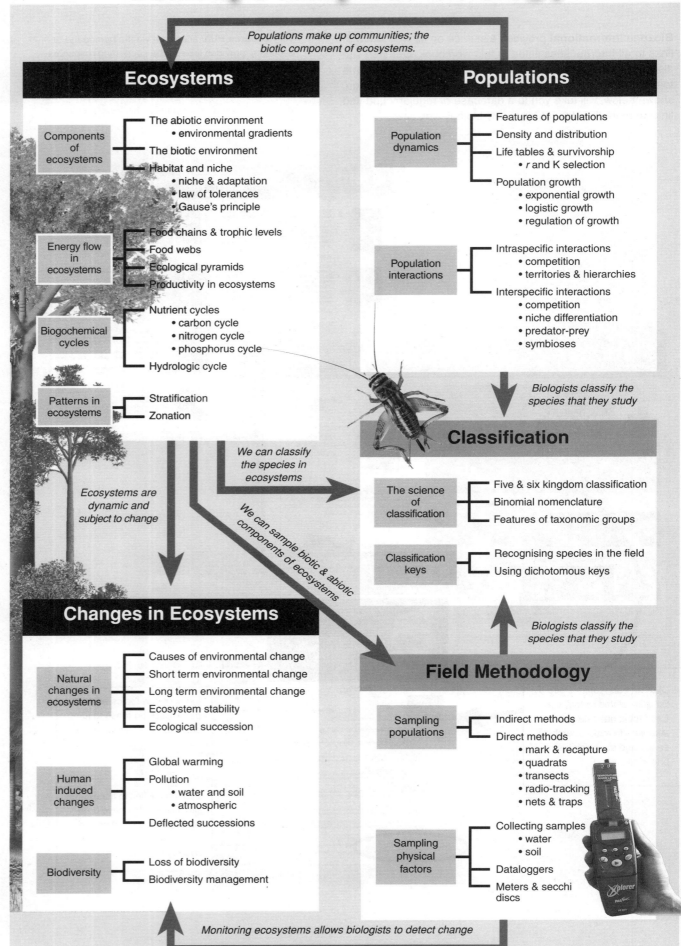

Populations make up communities; the biotic component of ecosystems.

Ecosystems

Components of ecosystems
- The abiotic environment
  - environmental gradients
- The biotic environment
- Habitat and niche
  - niche & adaptation
  - law of tolerances
  - Gause's principle

Energy flow in ecosystems
- Food chains & trophic levels
- Food webs
- Ecological pyramids
- Productivity in ecosystems

Biogochemical cycles
- Nutrient cycles
  - carbon cycle
  - nitrogen cycle
  - phosphorus cycle
- Hydrologic cycle

Patterns in ecosystems
- Stratification
- Zonation

Populations

Population dynamics
- Features of populations
- Density and distribution
- Life tables & survivorship
  - *r* and K selection
- Population growth
  - exponential growth
  - logistic growth
  - regulation of growth

Population interactions
- Intraspecific interactions
  - competition
  - territories & hierarchies
- Interspecific interactions
  - competition
  - niche differentiation
  - predator-prey
  - symbioses

Biologists classify the species that they study

We can classify the species in ecosystems

We can sample biotic & abiotic components of ecosystems

Ecosystems are dynamic and subject to change

Classification

The science of classification
- Five & six kingdom classification
- Binomial nomenclature
- Features of taxonomic groups

Classification keys
- Recognising species in the field
- Using dichotomous keys

Changes in Ecosystems

Natural changes in ecosystems
- Causes of environmental change
- Short term environmental change
- Long term environmental change
- Ecosystem stability
- Ecological succession

Human induced changes
- Global warming
- Pollution
  - water and soil
  - atmospheric
- Deflected successions

Biodiversity
- Loss of biodiversity
- Biodiversity management

Biologists classify the species that they study

Field Methodology

Sampling populations
- Indirect methods
- Direct methods
  - mark & recapture
  - quadrats
  - transects
  - radio-tracking
  - nets & traps

Sampling physical factors
- Collecting samples
  - water
  - soil
- Dataloggers
- Meters & secchi discs

Monitoring ecosystems allows biologists to detect change

Resources Information

Your set textbook should always be a starting point for information, but there are also many other resources available. A list of readily available resources is provided below. Access to the publishers of these resources can be made directly from Biozone's web site through our resources hub: **www.thebiozone.com/resource-hub.html**. Please note that our listing of any product in this workbook does not denote Biozone's endorsement of it.

Supplementary Texts

Reiss, M. & J. Chapman, 2000. **Environmental Biology**, 104 pp. **ISBN**: 0521787270
An introduction to environmental biology covering agriculture, pollution, resource conservation and conservation issues, and practical work in ecology. Questions and exercises are provided and each chapter includes an introduction and summary.

Adds, J., E. Larkcom, R. Miller, & R. Sutton, 1999. **Tools, Techniques and Assessment in Biology**, 160 pp. **ISBN**: 0-17-448273-6
A course guide covering basic lab protocols, microscopy, quantitative techniques in the lab and field, advanced DNA techniques and tissue culture, data handling and statistical tests, and exam preparation. Includes useful appendices.

Cadogan, A. and Ingram, M., 2002 **Maths for Advanced Biology**
Publisher: NelsonThornes
ISBN: 0-7487-6506-9
Comments: *This slim but comprehensive text covers the maths requirements for students at this level. Includes examples, and covers units, graphing and tabulation, descriptive statistics, and statistical tests for various types of data.*

Allen, D, M. Jones, and G. Williams, 2001. **Applied Ecology**, 104 pp. **ISBN**: 0-00-327741-0
Includes coverage of methods in practical ecology, the effects of pollution on diversity, adaptations, agricultural ecosystems and harvesting (including fisheries management), and conservation issues.

Smith, R.L. and T.M. Smith, R. 2001 (6 edn). **Ecology and Field Biology**, 700 pp. **ISBN**: 0321042905
A comprehensive overview of all aspects of ecology, including evolution, ecosystems theory, practical application, biogeochemical cycles, and global change. A field package, comprising a student "Ecology Action Guide" and a subscription to the web based "The Ecology Place" is also available.

Brower, J.E, J.H. Zar, & C.N. von Ende, 1997. **Field and Laboratory Methods for General Ecology**, 288 pp. (spiral bound)
Publisher: McGraw-Hill
ISBN: 0697243583
Comments: *An introductory manual for ecology, focussing on the collection, recording, and analysis of data. Provides balanced coverage of plant, animal, and physical elements.*

Biology Dictionaries

Access to a good biology dictionary is useful when dealing with biological terms. Some of the titles available are listed below. Link to the relevant publisher via Biozone's resources hub or by typing: **www.thebiozone.com/resources/dictionaries-pg1.html**

Clamp, A. **AS/A-Level Biology. Essential Word Dictionary**, 2000, 161 pp. Philip Allan Updates. **ISBN**: 0-86003-372-4.
Carefully selected essential words for AS and A2. Concise definitions are supported by further explanation and illustrations where required.

Hale, W.G., J.P. Margham, & V.A. Saunders. **Collins: Dictionary of Biology** 3 ed. 2003, 672 pp. HarperCollins. **ISBN**: 0-00-714709-0.
Updated to take in the latest developments in biology from the Human Genome Project to advancements in cloning (new edition pending).

Henderson, I.F, W.D. Henderson, and E. Lawrence. **Henderson's Dictionary of Biological Terms**, 1999, 736 pp. Prentice Hall. **ISBN**: 0582414989
This edition has been updated, rewritten for clarity, and reorganised for ease of use. An essential reference and the dictionary of choice for many.

McGraw-Hill (ed). **McGraw-Hill Dictionary of Bioscience**, 2 ed., 2002, 662 pp. McGraw-Hill. **ISBN**: 0-07-141043-0
22 000 entries encompassing more than 20 areas of the life sciences. It includes synonyms, acronyms, abbreviations, and pronunciations for all terms.

Periodicals, Magazines, & Journals

Biological Sciences Review: *An informative quarterly publication for biology students.* Enquiries: **UK**: Philip Allan Publishers **Tel**: 01869 338652 **Fax**: 01869 338803 **E-mail**: sales@philipallan.co.uk **Australasia**: **Tel**: 08 8278 5916, **E-mail**: rjmorton@adelaide.on.net

New Scientist: *Widely available weekly magazine with research summaries and features.* Enquiries: Reed Business Information Ltd, 51 Wardour St. London WIV 4BN **Tel**: (UK and intl):+44 (0) 1444 475636 **E-mail**: ns.subs@qss-uk.com *or subscribe from their web site.*

Scientific American: *A monthly magazine containing specialist features. Articles range in level of reading difficulty and assumed knowledge.* Subscription enquiries: 415 Madison Ave. New York. NY10017-1111 **Tel**: (outside North America): 515-247-7631 **Tel**: (US& Canada): 800-333-1199

School Science Review: *A quarterly journal which includes articles, reviews, and news on current research and curriculum development. Free to Ordinary Members of the ASE or available on subscription.* Enquiries: **Tel**: 01707 28300 **Email**: info@ase.org.uk *or visit their web site.*

The American Biology Teacher: *The peer-reviewed journal of the NABT. Published nine times a year and containing information and activities relevant to biology teachers.* Contact: NABT, 12030 Sunrise Valley Drive, #110, Reston, VA 20191-3409 **Web**: www.nabt.org

Ecosystems

Understanding ecosystems and investigating ecological patterns

Ecosystems and the influence of abiotic factors. Habitat, niche, and adaptation. Patterns of community distribution: zonation and stratification.

Learning Objectives

☐ 1. Compile your own glossary from the **KEY WORDS** displayed in **bold type** in the learning objectives below.

Biomes and Ecosystems *(pages 9-10)*

☐ 2. Define the terms: **ecology**, **ecosystem**, **community**, **population**, **species**, and **environment**. Recognise that the **biosphere** consists of interdependent ecosystems and that ecosystems are dynamic entities, subject to change.

☐ 3. Recognise major **biomes** on Earth and explain how they are classified according to major vegetation type. Appreciate the influence of latitude and local climate in determining the distribution of world biomes.

☐ 4. Identify the **biotic** and **abiotic** components of a named ecosystem. Include reference to the predominant **vegetation** type and the physical factors that determine the ecosystem's characteristics. Explain how the different parts of an ecosystem influence each other.

Habitats and Microclimates *(pages 11-18, 21-22)*

☐ 5. Explain how **environmental gradients** can occur over relatively short distances, e.g. on rocky shores, forests, lakes, or deserts. Explain how gradients in the abiotic environment contribute to community patterns.

☐ 6. Describe aspects of the **abiotic environment**, and explain how they influence the distribution of species in a local environment. Explain how **limiting factors** determine species distribution. Explain why limiting factors are often different for plants and animals.

☐ 7. List the factors commonly used to describe a **habitat**. Distinguish between habitat and **microhabitat** and explain how microhabitats arise as a result of differences in **microclimates**. Recognise the habitat as part of the **niche** of a **species**. Distinguish between the **tolerance range** and **optimum range** for a species.

Niche and Adaptation *(pages 23-27, 68-70)*

☐ 8. Describe the components of an organism's **ecological niche** (niche), and describe examples. Recognise the constraints that are normally placed on the niche occupied by an organism. Distinguish between the **fundamental** and the **realised niche**.

☐ 9. Understand the significance of Gause's **competitive exclusion** principle with respect to niche overlap between species. Appreciate the effect of **interspecific competition** on **niche breadth**.

☐ 10. Describe examples of **adaptive features (adaptations)** in named organisms. Recognise that organisms show **physiological**, **structural**, and **behavioural adaptations** for survival in a given niche and that these are the result of changes that occur to the species as a whole, but not to individuals within their own lifetimes.

Community Patterns *(pages 19-22)*

☐ 11. Recognise that community composition can change along an **environmental gradient**. Describe common types of distributional variation in communities:

(a) **Zonation**: Describe zonation in an intertidal community or altitudinal zonation in a forest community. Account for the distributional pattern in terms of the abiotic and biotic factors present.

(b) **Stratification**: Describe stratification in a forest community. Account for the distributional pattern in terms of the abiotic factors or biotic factors present.

See page 7 for additional details of these texts:
■ Allen, D, M. Jones, and G. Williams, 2001. **Applied Ecology**, chpt. 3.
■ Reiss, M. & J. Chapman, 2000. **Environmental Biology** (Cambridge University Press), pp. 1, 3, 76.
■ Smith, R. L. & T.M. Smith, 2001. **Ecology and Field Biology**, reading as required.

Presentation MEDIA to support this topic:
ECOLOGY
• Ecosystems • Niche
• Communities

See page 7 for details of publishers of periodicals:

STUDENT'S REFERENCE
■ **Big Weather** New Scientist, 22 May 1999 (Inside Science). *Global weather patterns and their role in shaping the ecology of the planet.*
■ **The Ecological Niche** Biol. Sci. Rev., 12(4), March 2000, pp. 31-35. *An excellent account of the niche - an often misunderstood concept that is never-the-less central to ecological theory.*
■ **Extreme Olympics** New Scientist, 30 March 2002 (Inside Science). *Extreme ecosystems and the adaptations of the organisms that live in them.*
■ **Cave Dwellers: Living without Light** Biol. Sci. Rev., 17(1) Sept. 2004, pp. 38-41. *Cave ecosystems and the adaptations of cave dwellers.*
■ **Biodiversity and Ecosystems** Biol. Sci. Rev., 11(4), March 1999, pp. 18-23. *Species richness and the breadth and overlap of niches. An account of how biodiversity influences ecosystem dynamics.*

See pages 4-5 for details of how to access **Bio Links** from our web site: **www.thebiozone.com**. From Bio Links, access sites under the topics:

Online Textbooks and Lecture Notes: > **Glossaries**: • Ecology glossary > **General Online Biology Resources**: • Ken's bio-web resources • Virtual library: bioscience
ECOLOGY: • Introduction to biogeography and ecology > **Ecosystems**: • Freshwater ecosystems

Biomes

Global patterns of vegetation distribution are closely related to climate. **Biomes** are large areas where the vegetation type shares a particular suite of physical requirements. They have characteristic features, but the boundaries between them are not distinct. The same biome may occur in widely separated regions of the world wherever the climatic and soil conditions are similar.

Wherever they occur, mountainous regions are associated with their own altitude adapted vegetation. The rainshadow effect of mountains governs the distribution of deserts in some areas too (e.g. in Chile). Although the classification of biomes may vary slightly, most classifications recognise desert, tundra, grassland and forest types and differentiate them on the basis of latitude.

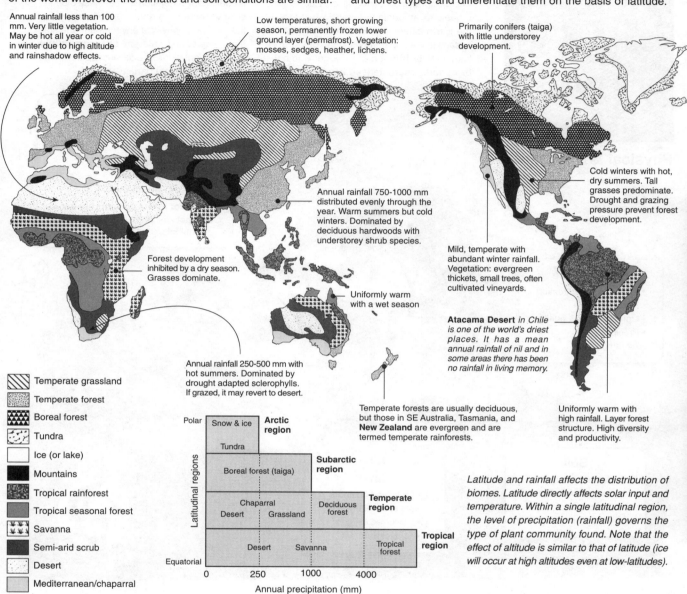

Annual rainfall less than 100 mm. Very little vegetation. May be hot all year or cold in winter due to high altitude and rainshadow effects.

Low temperatures, short growing season, permanently frozen lower ground layer (permafrost). Vegetation: mosses, sedges, heather, lichens.

Primarily conifers (taiga) with little understorey development.

Annual rainfall 750-1000 mm distributed evenly through the year. Warm summers but cold winters. Dominated by deciduous hardwoods with understorey shrub species.

Cold winters with hot, dry summers. Tall grasses predominate. Drought and grazing pressure prevent forest development.

Forest development inhibited by a dry season. Grasses dominate.

Mild, temperate with abundant winter rainfall. Vegetation: evergreen thickets, small trees, often cultivated vineyards.

Uniformly warm with a wet season

Atacama Desert in Chile is one of the world's driest places. It has a mean annual rainfall of nil and in some areas there has been no rainfall in living memory.

Annual rainfall 250-500 mm with hot summers. Dominated by drought adapted sclerophylls. If grazed, it may revert to desert.

Temperate forests are usually deciduous, but those in SE Australia, Tasmania, and **New Zealand** are evergreen and are termed temperate rainforests.

Uniformly warm with high rainfall. Layer forest structure. High diversity and productivity.

Legend:
- Temperate grassland
- Temperate forest
- Boreal forest
- Tundra
- Ice (or lake)
- Mountains
- Tropical rainforest
- Tropical seasonal forest
- Savanna
- Semi-arid scrub
- Desert
- Mediterranean/chaparral

Latitudinal regions (Polar to Equatorial) vs Annual precipitation (mm): 0, 250, 1000, 4000

- Polar: Snow & ice — **Arctic region**
- Tundra
- Boreal forest (taiga) — **Subarctic region**
- Chaparral / Desert / Grassland / Deciduous forest — **Temperate region**
- Desert / Savanna / Tropical forest — **Tropical region**
- Equatorial

Latitude and rainfall affects the distribution of biomes. Latitude directly affects solar input and temperature. Within a single latitudinal region, the level of precipitation (rainfall) governs the type of plant community found. Note that the effect of altitude is similar to that of latitude (ice will occur at high altitudes even at low-latitudes).

Ecosystems

1. Suggest what abiotic factor(s) limit the northern extent of boreal forest: _____

_____

2. Grasslands have about half the productivity of tropical rainforests, yet this is achieved with less than a tenth of the biomass; grasslands are more productive per unit of biomass. Suggest how this greater efficiency is achieved:

_____

_____

_____

3. Suggest a reason for the distribution of deserts and semi-desert areas in northern parts of Asia and in the west of North and South America (away from equatorial regions):

_____

_____

Code: A 2

Components of an Ecosystem

The concept of the ecosystem was developed to describe the way groups of organisms are predictably found together in their physical environment. A community comprises all the organisms within an ecosystem. Both physical (abiotic) and biotic factors affect the organisms in a community, influencing their distribution and their survival, growth, and reproduction.

The Biosphere

The **biosphere** containing all the Earth's living organisms amounts to a narrow belt around the Earth extending from the bottom of the oceans to the upper atmosphere. Broad scale life-zones or **biomes** are evident within the biosphere, characterised according to the predominant vegetation. Within these biomes, **ecosystems** form natural units comprising the non-living, physical environment (the soil, atmosphere, and water) and the **community** (all the populations of different species living and interacting in a particular area).

Physical Environment

Community: Biotic Factors
- Producers
- Consumers
- Detritivores
- Decomposers

Interact in the community as:
Competitors, parasites, pathogens, symbionts, predators, herbivores

Atmosphere
- Wind speed & direction
- Humidity
- Light intensity & quality
- Precipitation
- Air temperature

Soil
- Nutrient availability
- Soil moisture & pH
- Composition
- Temperature

Water
- Dissolved nutrients
- pH and salinity
- Dissolved oxygen
- Temperature

1. Distinguish clearly between a community and an ecosystem: _____

_____

_____

_____

2. Distinguish between biotic and abiotic factors: _____

_____

_____

3. Use one or more of the following terms to describe each of the features of a rainforest listed below:
   Terms: *population, community, ecosystem, physical factor.*

   (a) All the green tree frogs present: _____   (c) All the organisms present: _____

   (b) The entire forest: _____   (d) The humidity: _____

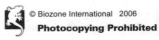

Code: A 1

Habitats

The environment in which an organism lives (including all the physical and biotic factors) is termed its **habitat**. For each of the organisms below, briefly describe their habitat (an example is provided). As well as describing the general environment (e.g. forest, ocean), include detail about the biotic and abiotic factors in the habitat which enable the organism to thrive.

Jew's ear fungus: *Auricularia auricula*

Herring gull: *Larus argentatus*

Coyote: *Canis latrans*

White-tailed deer: *Odocoileus virginianus*

Common barn owl: *Tyto alba*

Oak: *Quercus virginiana*

1. General habitat of the Jew's ear fungus: _Woodland, especially on elder._

   (a) Biotic factors: _Source of live and/or dead wood (usually elder) from which to obtain its nutrition._

   (b) Abiotic factors: _Autumnal needs cooler temperatures, high moisture levels in the soil and high humidity._

2. General habitat of the herring gull: _____

   (a) Biotic factors: _____

   (b) Abiotic factors: _____

3. General habitat of a coyote: _____

   (a) Biotic factors: _____

   (b) Abiotic factors: _____

4. General habitat of white-tailed deer: _____

   (a) Biotic factors: _____

   (b) Abiotic factors: _____

5. General habitat of the common barn owl: _____

   (a) Biotic factors: _____

   (b) Abiotic factors: _____

6. General habitat of an oak tree: _____

   (a) Biotic factors: _____

   (b) Abiotic factors: _____

Ecosystems

Code: RA 2

Law of Tolerances

The **law of tolerances** states that: *For each abiotic factor, an organism has a range of tolerances within which it can survive. Toward the upper and lower extremes of this tolerance range, that abiotic factor tends to limit the organism's ability to survive.* The wider an organism's tolerance range for a given abiotic factor (e.g. temperature, pH, salinity, turbidity, humidity, water pressure, light intensity), the more likely it is that the organism will be able to survive variations in that factor. Species **dispersal** is also strongly influenced by **tolerance range**. The wider the tolerance range of a species, the more widely dispersed the organism is likely to be. As well as a tolerance range, organisms have a narrower **optimum range** within which they function best. This may vary from one stage of an organism's development to another or from one season to another. Every species has its own optimum range. Organisms will usually be most abundant where the abiotic factors are closest to the optimum range.

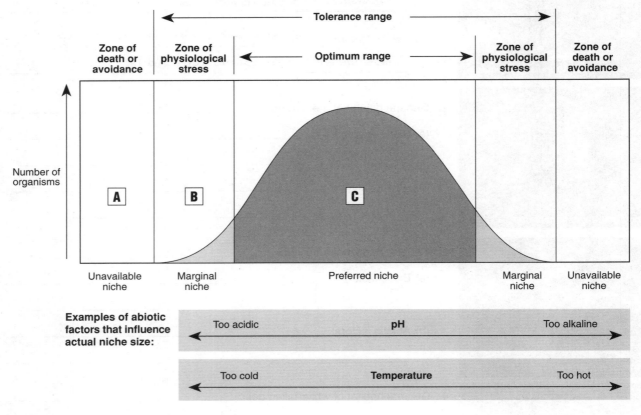

1. Study the diagram above and answer the following questions:

   (a) State the range in the diagram which contains the greatest number of organisms: _____

   _____

   (b) Explain why this is the case: _____

   _____

2. Organism C is occupying its preferred niche, in terms of physical factors. Explain what you would think would be the greatest constraint on organism C's growth and reproduction, within its preferred niche:

   _____

   _____

3. Organism B is occupying an area outside its preferred niche. Describe some probable stresses on B in this area:

   _____

   _____

   _____

4. Organism A has been forced, by crowding, to occupy an area completely unlike its preferred niche. Explain the likely result of organism A being forced into this area:

   _____

   _____

Dingo Habitats

An organism's habitat is not always of a single type. Some animals range over a variety of habitats, partly in order to obtain different resources from different habitats, and sometimes simply because they are forced into marginal habitats by competition. Dingoes are found throughout Australia, in ecosystems as diverse as the tropical rainforests of the north to the arid deserts of the Centre. Within each of these ecosystems, they may frequent several habitats or microhabitats. The information below shows how five dingo packs exploit a variety of habitats at one location in Australia. The table on the following page shows how dingoes are widespread in their distribution and are found living in a variety of ecosystems.

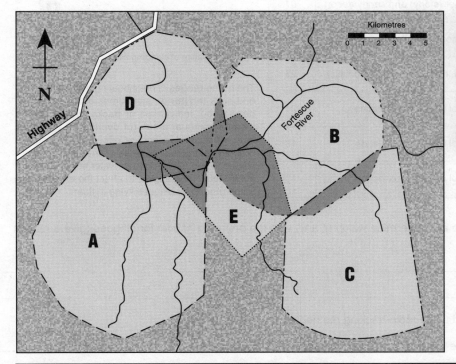

The map on the left shows the territories of five stable dingo packs (A to E) in the Fortescue River region in north-west Australia. The territories were determined by 4194 independent radio-tracking locations over 4 years. The size and nature of each territory, together with the makeup of each pack is given in the table below. The major prey of the dingoes in this region are large kangaroos (red kangaroos and euros).

Adapted from: Corbett, L. *The dingo in Australia and Asia*, 1995. University of NSW Press, after (original source) Thomson, P.C. 1992. *The behavioural ecology of dingoes in north-west Australia. IV. Social and spatial organisation, and movements. Wildlife Research* 19: 543-563.

Dingo pack name	Territory area (km²)	Pack size	Dingo density	Index of kangaroo abundance (%)	Habitat types (%) and habitat usage (%) in each territory							
					Riverine		Stony		Floodplain		Hills	
Pack A	113	12	10.6	15.9	10	(49)	1	(2)	21	(6)	69	(44)
Pack B	94	12		8.5	14	(43)	9	(10)	38	(25)	39	(23)
Pack C	86	3		3.9	2	(3)	0	(0)	63	(94)	35	(3)
Pack D	63	6		12.3	12	(35)	5	(8)	46	(20)	37	(37)
Pack E	45	10	22.2	8.4	14	(31)	6	(4)	39	(18)	42	(47)

mean number of dingoes per 100 km² (calculated by you)

Percentage of observations of kangaroos per observations of dingoes

Portion of the territory with this kind of habitat

Percentage of time spent by the pack in this habitat

1. Calculate the density of each of the dingo packs at the Fortescue River site above (two have been done for you). Remember that to determine the density, you carry out the following calculation:

   Density = pack size ÷ territory area x 100   (to give the mean number per 100 km²)

2. Name the dominant habitat (or habitats) for each territory in the table above (e.g. riverine, stony, floodplain, hills):

   (a) Pack A: _____

   (b) Pack B: _____

   (c) Pack C: _____

   (d) Pack D: _____

   (e) Pack E: _____

Ecosystems

Code: DA 2

14

Dingo home range size in contrasting ecosystems

Location (study site)	Ecosystem	Range (km²)
1 Fortescue River, North-west Australia	Semi-arid, coastal plains and hills	77
2 Simpson Desert, Central Australia	Arid, gibber (stony) and sandy desert	67
3 Kapalga, Kakadu N.P., North Australia	Tropical, coastal wetlands and forests	39
4 Harts Ranges, Central Australia	Semi-arid, river catchment and hills	25
5 Kosciusko N.P., South-east Australia	Moist, cool forested mountains	21
6 Georges Creek N.R., East Australia	Moist, cool forested tablelands	18
7 Nadgee N.R., South-east Australia	Moist, cool coastal forests	10

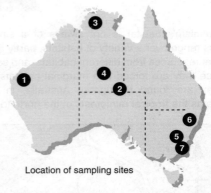

Location of sampling sites

The **home ranges** of neighbouring dingo individuals and pack **territories** sometimes overlap to some degree, but individuals or packs avoid being in these communal areas (of overlap) at the same time. This overlap occurs especially during the breeding season or in areas where resources are shared (e.g. hunting grounds, water holes). Different ecosystem types appear to affect the extent of the home ranges for dingoes living in them.

3. Study the table on the previous page and determine which (if any) was the preferred habitat for dingoes (give a reason for your answer):

4. The dingoes at this site were studied using radio-tracking methods.

 (a) Explain how radio-tracking can be used to determine the movements of dingoes:

 (b) State how many independent tracking locations were recorded during this study:

 (c) State how long a period of time the study was run for:

 (d) Explain why so many records were needed over such a long period of time:

5. From the table on the previous page, state whether the relative kangaroo abundance (the major prey of the dingo) affects the density that a given territory can support:

6. Study the table at the top of this page which shows the results of an investigation into the sizes of home ranges in dingo populations from different ecosystems.

 (a) Describe the feature of the ecosystems that appears to affect the size of home ranges:

 (b) Explain how this feature might affect how diverse (varied) the habitats are within the ecosystem:

Physical Factors and Gradients

Gradients in abiotic factors are found in almost every environment; they influence habitats and microclimates, and determine patterns of species distribution. This activity, covering the next four pages, examines the physical gradients and microclimates that might typically be found in four, very different environments. Note that **dataloggers** (pictured right), are being increasingly used to gather such data. The principles of their use are covered in the topic *Practical Ecology*.

A Desert Environment

Desert environments experience extremes in temperature and humidity, but they are not uniform with respect to these factors. This diagram illustrates hypothetical values for temperature and humidity for some of the microclimates found in a desert environment at midday.

300 m
altitude

Burrow	**Under rock**	**Surface**	**Crevice**	**High air**	**Low air**
25°C	28°C	45°C	27°C	27°C	33°C
95% Hum	60% Hum	<20% Hum	95% Hum	20% Hum	20% Hum

1 m above
the ground

1 m underground

2 m underground

Ecosystems

1. Distinguish between **climate** and **microclimate**: _____

_____

_____

2. Study the diagram above and describe the general conditions where high humidity is found: _____

_____

_____

3. Identify the three microclimates that a land animal might exploit to avoid the extreme high temperatures of midday:

_____

_____

4. Describe the likely consequences for an animal that was unable to find a suitable microclimate to escape midday sun:

_____

_____

5. Describe the advantage of high humidity to the survival of most land animals: _____

_____

_____

6. Describe the likely changes to the temperature and relative humidity that occur during the night: _____

_____

_____

Code: DA 2

Physical Factors in a Tropical Rainforest

Canopy

Light: 70%
Wind: 15 kmh^{-1}
Humid: 67%

Light: 50%
Wind: 12 kmh^{-1}
Humid: 75%

Light: 12%
Wind: 9 kmh^{-1}
Humid: 80%

Light: 6%
Wind: 5 kmh^{-1}
Humid: 85%

Light: 1%
Wind: 3 kmh^{-1}
Humid: 90%

Light: 0%
Wind: 0 kmh^{-1}
Humid: 98%

A **datalogger** fitted with suitable probes was used to gather data on wind speed (**Wind**), humidity (**Humid**), and light intensity (**Light**) for each layer (left). Light intensity is given as a percentage of full sunlight.

Leaf litter

Tropical rainforests are complex communities with a vertical structure which divides the vegetation into layers. This pattern of vertical layering is called **stratification**.

7. Describe the environmental gradient (general trend) from the canopy to the leaf litter for:

   (a) Light intensity: _____

   (b) Wind speed: _____

   (c) Humidity: _____

8. Explain why each of these factors changes as the distance from the canopy increases:

   (a) Light intensity: _____

   _____

   (b) Wind speed: _____

   _____

   (c) Humidity: _____

   _____

9. Apart from the light intensity, describe the other feature of light that will change with distance from the canopy:

   _____

   _____

10. Plants growing on the forest floor have some advantages and disadvantages with respect to the physical factors.

    (a) Describe one advantage: _____

    (b) Describe one disadvantage: _____

Physical Factors at Low Tide on a Rock Platform

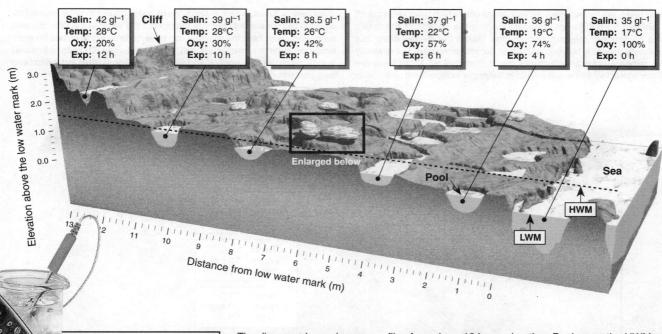

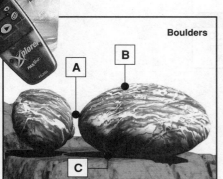

Boulders

The diagram above shows a profile of a rock platform at low tide. The **high water mark** (HWM) shown here is the average height the spring tide rises to. In reality, the high tide level will vary with the phases of the moon (i.e. spring tides and neap tides). The **low water mark** (LWM) is an average level subject to the same variations due to the lunar cycle. The rock pools vary in size, depth, and position on the platform. They are isolated at different elevations, trapping water from the ocean for time periods that may be brief or up to 10 –

12 hours duration. Pools near the HWM are exposed for longer periods of time than those near the LWM. The difference in exposure times results in some of the physical factors exhibiting a **gradient**; the factor's value gradually changes over distance. Physical factors sampled in the pools include salinity, or the amount of dissolved salts (g) per liter (**Salin**), temperature (**Temp**), dissolved oxygen compared to that of open ocean water (**Oxy**), and exposure, or the amount of time isolated from the ocean water (**Exp**).

11. Describe the environmental gradient (general trend) from the low water mark (LWM) to the high water mark (HWM) for:

(a) Salinity: _____

(b) Temperature: _____

(c) Dissolved oxygen: _____

(d) Exposure: _____

12. Rock pools above the normal high water mark (HWM), such as the uppermost pool in the diagram above, can have wide extremes of salinity. Explain the conditions under which these pools might have either:

(a) Very low salinity: _____

_____

(b) Very high salinity: _____

_____

13. (a) The inset diagram (above, left) is an enlarged view of two boulders on the rock platform. Describe how the physical factors listed below might differ at each of the labelled points **A**, **B**, and **C**:

Mechanical force of wave action: _____

_____

Surface temperature when exposed: _____

_____

(b) State the term given to these localised variations in physical conditions: _____

Ecosystems

Physical Factors in an Oxbow Lake in Summer

Oxbow lakes are formed from old river meanders which have been cut off and isolated from the main channel following the change of the river's course. Commonly, they are very shallow (about 2-4 metres deep) but occasionally they may be deep enough to develop temporary, but relatively stable, temperature gradients from top to bottom (below). Small lakes are relatively closed systems and events in them are independent of those in other nearby lakes, where quite different water quality may be found. The physical factors are not constant throughout the water in the lake. Surface water and water near the margins can have quite different values for such factors as water temperature (**Temp**), dissolved oxygen (**Oxygen**) measured in milligrams per litre (mgl^{-1}), and light penetration (**Light**) indicated here as a percentage of the light striking the surface.

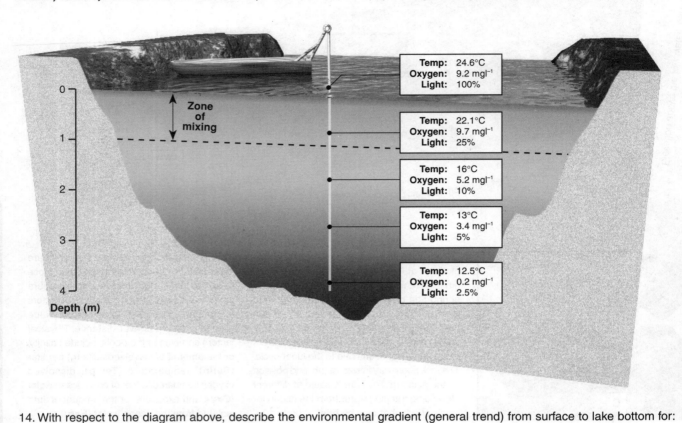

14. With respect to the diagram above, describe the environmental gradient (general trend) from surface to lake bottom for:

    (a) Water temperature: _____

    (b) Dissolved oxygen: _____

    (c) Light penetration: _____

15. During the summer months, the warm surface waters are mixed by gentle wind action. Deeper cool waters are isolated from this surface water. This sudden change in the temperature profile is called a **thermocline** which itself is a further barrier to the mixing of shallow and deeper water.

    (a) Explain the effect of the thermocline on the dissolved oxygen at the bottom of the lake: _____

    _____

    _____

    (b) Explain what causes the oxygen level to drop to the low level: _____

    _____

16. Many of these shallow lakes can undergo great changes in their salinity (sodium, magnesium, and calcium chlorides):

    (a) Name an event that could suddenly reduce the salinity of a small lake: _____

    (b) Name a process that can gradually increase the salinity of a small lake: _____

17. Describe the general effect of physical gradients on the distribution of organisms in habitats: _____

    _____

    _____

    _____

Community Change With Altitude

The Kosciusko National Park lies on the border between Victoria and New South Wales in Australia. In 1959, a researcher by the name of A.B. Costin carried out a detailed sampling of a transect between Berridale and the summit of Mt. Kosciusko. The map on the right shows the transect as a dotted line representing a distance of some 50-60 km.

The distribution of the various plant species up the slope of Mount Kosciusko is affected by changes in the physical conditions with increasing altitude. Below are two diagrams, one showing the profile of the transect showing changes in vegetation and soil types with increasing altitude. The diagram below the profile shows the changes in temperature and rainfall (precipitation) with altitude.

The low altitude soil around Berridale has low levels of organic matter supporting dry tussock grassland vegetation. The high altitude alpine soils are rich in organic matter, largely due to slow decay rates.

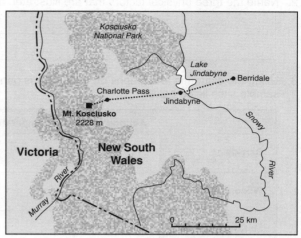

Ecosystems

Profile of Mount Kosciusko

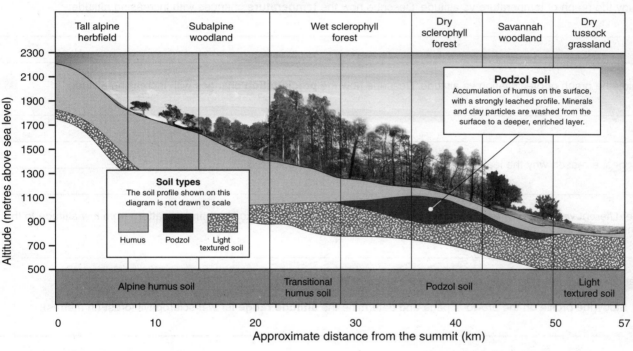

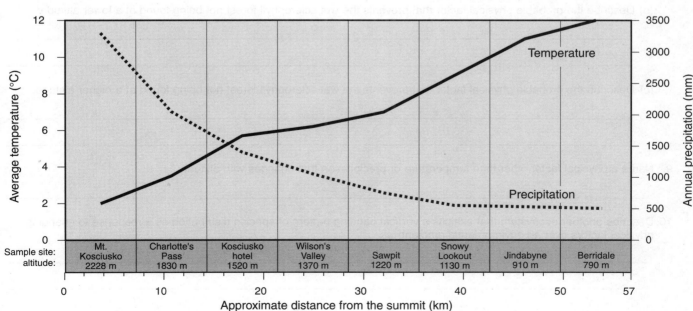

1. Calculate the vertical distance (change in altitude) in metres, between Berridale and Mount Kosciusko: _____

2. Name the three physical factors that are illustrated on the diagrams on the previous page:

   _____

   _____

3. Using the diagrams and graphs on the previous page, describe the following physical measurements for the three sample sites listed below:

	Altitude (m)	Temperature (°C)	Precipitation (mm)	Soil type
Berridale:				
Wilson's Valley:				
Mt Kosciusko:				

4. Study the graph of temperature vs altitude. Describe how the **temperature** changes with increasing altitude:

   _____

   _____

5. Study the graph of precipitation vs altitude. Describe how the **precipitation** changes with increasing altitude:

   _____

   _____

6. Suggest a reason why the leaf litter is slow to decay in the alpine soil: _____

   _____

7. The different vegetation types are distributed on the slopes of Mt. Kosciusko in a banded pattern from low altitude to the summit. Give the name of this kind of distribution pattern:

   _____

8. Wet sclerophyll forest is found part way up the slope of Mt. Kosciusko.

   (a) Study the profile on the previous page and determine the **altitude range** for wet sclerophyll forest (in metres):

   _____

   (b) Describe the probable physical factor that prevents the wet sclerophyll forest not being found at a lower altitude:

   _____

   _____

   (c) Describe the probable physical factor that prevents the wet sclerophyll forest not being found at a higher altitude:

   _____

   _____

9. Name a physical factor other than temperature or precipitation that changes with altitude:

   _____

10. Describe another ecosystem that exhibits a vertical banding pattern of species distribution as a response to changing physical factors over an environmental gradient:

   _____

   _____

   _____

Shoreline Zonation

Zonation refers to the division of an ecosystem into distinct zones that experience similar abiotic conditions. In a more global sense, differences in latitude and altitude create distinctive zones of vegetation type, or **biomes**. Zonation is particularly clear on a rocky seashore, where assemblages of different species form a banding pattern approximately parallel to the waterline. This effect is marked in temperate regions where the prevailing weather comes from the same general direction. Exposed shores show the clearest zonation. On sheltered rocky shores there is considerable species overlap and it is only on the upper shore that distinct zones are evident. Rocky shores exist where wave action prevents the deposition of much sediment. The rock forms a stable platform for the secure attachment of organisms such as large seaweeds and barnacles. Sandy shores are less stable than rocky shores and the organisms found there are adapted to the more mobile substrate.

Rocky shore at Sleahead, Ireland.

Seashore Zonation Patterns

The zonation of species distribution according to an environmental gradient is well shown on rocky shorelines. In Britain, exposed rocky shores occur along much of the western coasts. Variations in low and high tide affect zonation, and in areas with little tidal variation, zonation is restricted. High on the shore, some organisms may be submerged only at spring high tide. Low on the shore, others may be exposed only at spring low tide. There is a gradation in extent of exposure and the physical conditions associated with this. Zonation patterns generally reflect the vertical movement of seawater. Sheer rocks can show marked zonation as a result of tidal changes with little or no horizontal shift in species distribution. The profiles below left, show zonation patterns on an exposed rocky shore (left profile) with an exposed sandy shore for comparison (right profile). **SLT** = spring low tide mark, **MLT** = mean low tide mark, **MHT** = mean high tide mark, **SHT** = spring high tide mark.

Key to species

1. Lichen: sea ivory
2. Small periwinkle *Littorina neritoides*
3. Lichen *Verrucaria maura*
4. Rough periwinkle *Littorina saxatilis*
5. Common limpet *Patella vulgaris*
6. Laver *Porphyra*
7. Spiral wrack *Fucus spiralis*
8. Australian barnacle
9. Common mussel *Mytilus edulis*
10. Common whelk *Buccinum undatum*
11. Grey topshell *Gibbula cineraria*
12. Carrageen (Irish moss) *Chondrus crispus*
13. Thongweed *Himanthalia elongata*
14. Toothed wrack *Fucus serratus*
15. Dabberlocks *Alaria esculenta*
16. Common sandhopper
17. Sandhopper *Bathyporeia pelagica*
18. Common cockle *Cerastoderma edule*
19. Lugworm *Arenicola marina*
20. Sting winkle *Ocinebra erinacea*
21. Common necklace shell *Natica alderi*
22. Rayed trough shell *Mactra corallina*
23. Sand mason worm *Lanice conchilega*
24. Sea anemone *Halcampa*
25. Pod razor shell *Ensis siliqua*
26. Sea potato *Echinocardium* (a heart urchin)

Note: Where several species are indicated within a single zonal band, they occupy the entire zone, not just the position where their number appears.

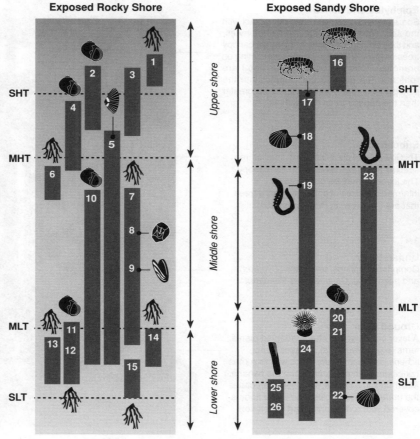

Exposed Rocky Shore **Exposed Sandy Shore**

1. (a) Suggest why the time of exposure above water is a major factor controlling species distribution on a rocky shore:

_____

(b) Identify two other abiotic factors that might influence species distribution on a rocky shore: _____

_____

(c) Identify two biotic factors that might influence species distribution on a rocky shore: _____

_____

2. Describe the zonation pattern on a rocky shore: _____

_____

Ecosystems

Code: A 2

Stratification in a Forest

Forest communities throughout the world have a structure that divides them into layers of different vegetation types. Tropical rainforests are complex and can be divided into four layers that represent zones of different vegetation. This vertical layering is called **stratification**. Perching plants, called epiphytes, grow in the forks of branching limbs in the canopy trees. The enormous diversity of species found in rainforests can be supported because of the wide variety of microhabitats provided by the layered structure of the forest. Species composition varies from region to region according to the altitude, rainfall, light levels, soil type, and drainage. Even the past history of the area can have an effect. The tallest trees are the emergents and may be 60 metres in height. Canopy trees fill the gaps between the emergents. Large woody climbing vines called lianes are found interwoven between the canopy trees. Below the canopy layer spreads the subcanopy of saplings, tree ferns and palms.

Structure of a Tropical Rainforest

Canopy
The topmost canopy layer is usually uneven, reaching a height of 20-40 metres and intercepting most of the direct sunlight falling on the forest. In places there are scattered broad-leaved emergents, with the sides of the crowns exposed above the canopy. Species include walnuts (*Beilschmiedia* spp.), cassowary satinash, and banana fig.

Epiphytes and lianes
Some ferns and orchids have no contact with the soil. These plants are called epiphytes, and they cling to the trunks of the larger canopy trees or grow in the leaf litter accumulating on branches. Lianes (e.g. *Calamus* spp.) are rooted in the ground, but clamber into the canopy. They may develop a leaf area greater than the canopy trees themselves.

Subcanopy
This lower level contains some smaller tree species that may never reach the canopy. The layering may be obscured by intermediate sizes of potential canopy trees that are still growing (transgressives).

Understorey
Comprises saplings, tree ferns (*Cyathea*), and palms (e.g. *Calamus* and *Pandanus*).

Ground layer
A layer composed of shade adapted plants: ferns (e.g. *Selaginella*), mosses, fungi, lichens, and giant fleshy herbs with rhizomes (zingibers such as *Alphinia* and the banana, *Musa*). In some forests, this layer incorporates the understorey species. The forest floor is covered with rotting and dead vegetation.

1. Draw horizontal lines on the diagram to indicate the boundary of each stratification layer. Label each layer.

2. Review the page on physical gradients in a rainforest before answering this question:

    (a) Describe some of the general adaptations to physical conditions in the upper forest shown by canopy trees:

    _____

    _____

    (b) Describe some of the general adaptations to physical conditions shown by plants growing on the forest floor:

    _____

    _____

Ecological Niche

The concept of the ecological niche has been variously described as an organism's 'job' or 'profession'. This is rather too simplistic for senior biology level. The **ecological niche** is better described as the functional position of an organism in its environment, comprising its habitat and the resources it obtains there, and the periods of time during which it is active. The diagram below illustrates the components that together define the niche of any organism. The full range of environmental conditions (biological and physical) under which an organism can exist describes its **fundamental niche**. As a result of pressure from, and interactions with, other organisms (e.g. superior competitors) species are usually forced to occupy a niche that is narrower than this and to which they are most highly adapted. This is termed the **realised niche**.

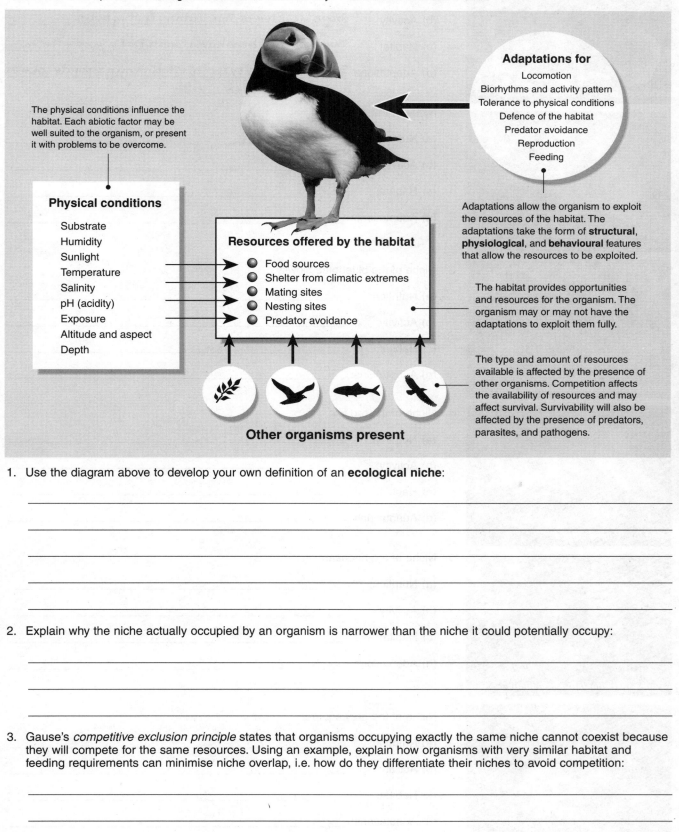

The physical conditions influence the habitat. Each abiotic factor may be well suited to the organism, or present it with problems to be overcome.

Adaptations for
Locomotion
Biorhythms and activity pattern
Tolerance to physical conditions
Defence of the habitat
Predator avoidance
Reproduction
Feeding

Physical conditions
Substrate
Humidity
Sunlight
Temperature
Salinity
pH (acidity)
Exposure
Altitude and aspect
Depth

Resources offered by the habitat
● Food sources
● Shelter from climatic extremes
● Mating sites
● Nesting sites
● Predator avoidance

Adaptations allow the organism to exploit the resources of the habitat. The adaptations take the form of **structural**, **physiological**, and **behavioural** features that allow the resources to be exploited.

The habitat provides opportunities and resources for the organism. The organism may or may not have the adaptations to exploit them fully.

The type and amount of resources available is affected by the presence of other organisms. Competition affects the availability of resources and may affect survival. Survivability will also be affected by the presence of predators, parasites, and pathogens.

Other organisms present

Ecosystems

1. Use the diagram above to develop your own definition of an **ecological niche**:

_____

_____

_____

_____

_____

2. Explain why the niche actually occupied by an organism is narrower than the niche it could potentially occupy:

_____

_____

_____

3. Gause's *competitive exclusion principle* states that organisms occupying exactly the same niche cannot coexist because they will compete for the same resources. Using an example, explain how organisms with very similar habitat and feeding requirements can minimise niche overlap, i.e. how do they differentiate their niches to avoid competition:

_____

_____

_____

Code: A 2

Ecological Niches

The concept of an organism's **ecological niche** is fundamental to understanding ecology. To fully describe a niche, take note of an organism's: trophic level, mode of feeding, activity periods, habitat and the resources exploited within it, as well as the adaptive features they possess to exploit them. Describe the **niche** for each of the following organisms:

White rot bracket fungus: *Coriolus*

1. Niche of the white rot bracket fungus:

   (a) Nutrition: Parasite of broadleaved trees using extracellular digestion.

   (b) Activity: Grows in clusters of thin, leathery fruiting bodies.

   (c) Habitat: Throughout most broad leaved forests. Pest in wooden fencing.

   (d) Adaptations: Spreads vigorously by spores, establishing as separate colonies

Sea star: *Astropecten*

2. Niche of the sea star:

   (a) Nutrition: _____

   (b) Activity: _____

   (c) Habitat: _____

   (d) Adaptations: _____

Bluebottle blowfly: *Calliphora vomitoria*

3. Niche of the bluebottle blowfly:

   (a) Nutrition: _____

   (b) Activity: _____

   (c) Habitat: _____

   (d) Adaptations: _____

Red fox: *Vulpes vulpes*

4. Niche of the red fox:

   (a) Nutrition: _____

   (b) Activity: _____

   (c) Habitat: _____

   (d) Adaptations: _____

Pheasant: *Phasianus colchicus*

5. Niche of the pheasant:

   (a) Nutrition: _____

   (b) Activity: _____

   (c) Habitat: _____

   (d) Adaptations: _____

Grey squirrel: *Sciurus carolinensis*

6. Niche of the grey squirrel:

   (a) Nutrition: _____

   (b) Activity: _____

   (c) Habitat: _____

   (d) Adaptations: _____

Adaptations to Niche

The adaptive features that evolve in species are the result of selection pressures on them through the course of their evolution. These features enable an organism to function most effectively in its niche, enhancing its exploitation of its environment and therefore its survival. The examples below illustrate some of the adaptations of two species: a British placental mammal and a migratory Arctic bird. Note that adaptations may be associated with an animal's structure (morphology), its internal physiology, or its behaviour.

Northern or Common Mole
(Talpa europaea)

Head-body length: 113-159 mm, tail length: 25-40 mm, weight range: 70-130 g.

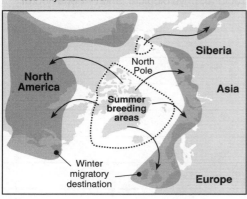

Mole hill

Lining of dry grass

Adult

Young

Moles (photos above) spend most of the time underground and are rarely seen at the surface. Mole hills are the piles of soil excavated from the tunnels and pushed to the surface. The cutaway view above shows a section of tunnels and a nest chamber. Nests are used for sleeping and raising young. They are dug out within the tunnel system and lined with dry plant material.

The northern (common) mole is a widespread insectivore found throughout most of Britain and Europe, apart from Ireland. They are found in most habitats but are less common in coniferous forest, moorland, and sand dunes, where their prey (earthworms and insect larvae) are rare. They are well adapted to life underground and burrow extensively, using enlarged forefeet for digging. Their small size, tubular body shape, and heavily buttressed head and neck are typical of burrowing species.

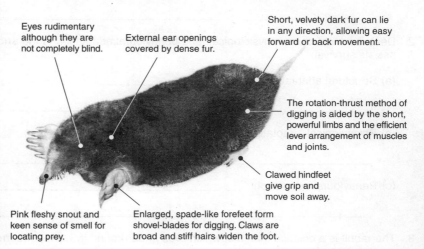

Eyes rudimentary although they are not completely blind.

External ear openings covered by dense fur.

Short, velvety dark fur can lie in any direction, allowing easy forward or back movement.

The rotation-thrust method of digging is aided by the short, powerful limbs and the efficient lever arrangement of muscles and joints.

Clawed hindfeet give grip and move soil away.

Pink fleshy snout and keen sense of smell for locating prey.

Enlarged, spade-like forefeet form shovel-blades for digging. Claws are broad and stiff hairs widen the foot.

Habitat and ecology: Moles spend most of their lives in underground tunnels. Surface tunnels occur where their prey is concentrated at the surface (e.g. land under cultivation). Deeper, permanent tunnels form a complex network used repeatedly for feeding and nesting, sometimes for several generations. **Senses and behaviour**: Keen sense of smell but almost blind. Both sexes are solitary and territorial except during breeding. Life span about 3 years. Moles are prey for owls, buzzards, stoats, cats, and dogs. Their activities aerate the soil and they control many soil pests. Despite this, they are regularly trapped and poisoned as pests.

Snow Bunting
(Plectrophenax nivalis)

The snow bunting is a small ground feeding bird that lives and breeds in the Arctic and sub-Arctic islands. Although migratory, snow buntings do not move to traditional winter homes but prefer winter habitats that resemble their Arctic breeding grounds, such as bleak shores or open fields of northern Britain and the eastern United States.

Snow buntings have the unique ability to moult very rapidly after breeding. During the warmer months, the buntings are a brown colour, changing to white in winter (right). They must complete this colour change quickly, so that they have a new set of feathers before the onset of winter and before migration. In order to achieve this, snow buntings lose as many as four or five of their main flight wing feathers at once, as opposed to most birds, which lose only one or two.

Very few small birds breed in the Arctic, because most small birds lose more heat than larger ones. In addition, birds that breed in the brief Arctic summer must migrate before the onset of winter, often travelling over large expanses of water. Large, long winged birds are better able to do this. However, the snow bunting is superbly adapted to survive in the extreme cold of the Arctic region.

White feathers are hollow and filled with air, which acts as an insulator. In the dark coloured feathers the internal spaces are filled with pigmented cells.

Less heat is lost from white plumage compared to dark plumage.

Snow buntings, on average, lay one or two more eggs than equivalent species further south. They are able to rear more young because the continuous daylight and the abundance of insects at high latitudes enables them to feed their chicks around the clock.

During snow storms or periods of high wind, snow buntings will burrow into snowdrifts for shelter.

Siberia

North Pole

North America

Asia

Summer breeding areas

Winter migratory destination

Europe

Habitat and ecology: Widespread throughout Arctic and sub-Arctic Islands. Active throughout the day and night, resting for only 2-3 hours in any 24 hour period. Snow buntings may migrate up to 6000 km but are always found at high latitudes. **Reproduction and behaviour**: The nest, which is concealed amongst stones, is made from dead grass, moss, and lichen. The male bird feeds his mate during the incubation period and helps to feed the young.

Ecosystems

Code: RA 2

1. Describe a structural, physiological, and behavioural adaptation of the **common mole**, explaining how each adaptation assists survival:

   (a) Structural adaptation: _____

   _____

   (b) Physiological adaptation: _____

   _____

   (c) Behavioural adaptation: _____

   _____

2. Describe a structural, physiological, and behavioural adaptation of the **snow bunting**, explaining how each adaptation assists survival:

   (a) Structural adaptation: _____

   _____

   (b) Physiological adaptation: _____

   _____

   (c) Behavioural adaptation: _____

   _____

3. The rabbit is a colonial mammal which lives underground in warrens (burrow systems) and feeds on grasses, cereal crops, roots, and young trees. Rabbits are a hugely successful species worldwide and often reach plague proportions. Through discussion, or your own knowledge and research, describe **six adaptations** of rabbits, identifying them as structural (S), physiological (P), or behavioural (B). The examples below are typical:

   Structural: Widely spaced eyes gives wide field of vision for surveillance and detection of danger.

   Physiological: High reproductive rate; short gestation and high fertility aids rapid population increases when food is available.

   Behavioural: Freeze behaviour when startled reduces the possibility of detection by wandering predators.

   (a) _____

   (b) _____

   (c) _____

   (d) _____

   (e) _____

   (f) _____

4. Examples of adaptations are listed below. Identify them as predominantly structural, physiological, and/or behavioural:

   (a) Relationship of body size and shape to latitude (tropical or Arctic): _____

   (b) The production of concentrated urine in desert dwelling mammals: _____

   (c) The summer and winter migratory patterns in birds and mammals: _____

   (d) The C4 photosynthetic pathway and CAM metabolism of plants: _____

   (e) The thick leaves and sunken stomata of desert plants: _____

   (f) Hibernation or torpor in small mammals over winter: _____

   (g) Basking in lizards and snakes: _____

Competition and Niche Size

Niche size is affected by competition. The effect on niche size will vary depending on whether the competition is weak, moderate or intense, and whether it is between members of the same species (**intraspecific**) or between different species (**interspecific**). The theoretical effects of competition on niche size are outlined in the diagram below.

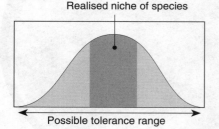
Realised niche of species

Narrow niche

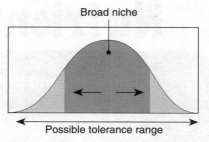
Broad niche

Possible tolerance range

Possible tolerance range

Possible tolerance range

Moderate interspecific competition

The tolerance range represents the potential (**fundamental**) niche a species could exploit. The actual or **realised** niche of a species is narrower than this because of competition. Niches of closely related species may overlap at the extremes, resulting in competition for resources in the zones of overlap.

Intense interspecific competition

When the competition from one or more closely related species becomes intense, there is selection for a more limited niche. This severe competition prevents a species from exploiting potential resources in the more extreme parts of its tolerance range. As a result, niche breadth decreases (the niche becomes narrower).

Intense intraspecific competition

Competition is most severe between individuals of the same species, because their resource requirements are usually identical. When intraspecific competition is intense, individuals are forced to exploit resources in the extremes of their tolerance range. This leads to expansion of the realised niche to less preferred areas.

Overlap in resource use between competing species

From the concept of the niche arose the idea that two species with the same niche requirements could not coexist, because they would compete for the same resources, and one would exclude the other. This is known as Gause's "**competitive exclusion principle**". If two species compete for some of the same resources (e.g. food items of a particular size), their resource use curves will overlap. Within the zone of overlap competition between the two species will be intense.

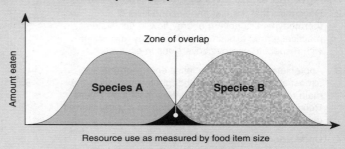

1. Distinguish between interspecific competition and intraspecific competition, and contrast the effect of these two types of competition on niche breadth:

   _____

   _____

   _____

   _____

2. Study the diagram above illustrating niche overlap between competing species and then answer the following questions:

   (a) Explain what you would expect to happen if the overlap in resource use of species **A** and **B** was to become very large (i.e. they utilised almost the same sized food item):

   _____

   (b) Describe how the degree of resource overlap might change seasonally: _____

   _____

   (c) If the zone of overlap between the resource use of the two species increased a little more from what is shown on the diagram, explain what is likely to happen to the breadth of the **realised** niche of each species:

   _____

3. Niche breadth can become broader in the presence of intense competition between members of the same species (diagram, top right). Describe one other reason why niche breadth could be very wide:

   _____

   _____

Ecosystems

Code: DA 2

Energy Flow & Nutrient Cycles

Energy flow and biogeochemical cycles

Energy flow in ecosystems: trophic relationships, ecological pyramids and productivity. Biogeochemical cycles.

Learning Objectives

☐ 1. Compile your own glossary from the **KEY WORDS** displayed in **bold type** in the learning objectives below.

Energy in Ecosystems

Background to the laws of energy *(page 30)*

☐ 2. Recognise that the first and second laws of thermodynamics govern energy flow in ecosystems. In practical terms, explain what the energy laws mean with respect to energy conversions in ecosystems.

☐ 3. Appreciate that light is the initial energy source for almost all ecosystems and that photosynthesis is the main route by which energy enters most food chains. Recognise that energy is dissipated as it is transferred through trophic levels.

☐ 4. Understand the interrelationship between nutrient cycling and energy flow in ecosystems:
  • Energy flows through ecosystems in the high energy chemical bonds within **organic matter**.
  • Nutrients move within and between ecosystems in **biogeochemical cycles** involving exchanges between the atmosphere, the Earth's crust, water, and living organisms.

Trophic relationships *(pages 33-40)*

☐ 5. Describe how the energy flow in ecosystems is described using **trophic levels**. Distinguish between **producers**, **primary and secondary consumers**, **detritivores**, and **saprotrophs (decomposers)**, identifying the relationship between them. Describe the role of each of these in energy transfer.

☐ 6. Describe how energy is transferred between different trophic levels in **food chains** and **food webs**. Compare the amount of energy available to each trophic level and recognise that the **efficiency** of this energy transfer is only 10-20%. Explain what happens to the remaining energy and why energy cannot be recycled.

☐ 7. With respect to the **efficiency** of energy transfer in food chains, explain the small biomass and low numbers of organisms at higher trophic levels.

☐ 8. Provide three examples of food chains, each with at least three linkages. Show how the (named) organisms are interconnected through their feeding relationships. Assign trophic levels to the organisms in a food chain.

☐ 9. Construct a food web for a named community, containing up to 10 organisms, showing how the (named) organisms are interconnected through their feeding relationships. Assign trophic levels to the organisms in the food web.

☐ 10. Recognise and discuss the difficulties of classifying organisms into trophic levels.

☐ 11. Explain what is meant **bioaccumulation** and describe an example. In your explanation, make it clear that you understand the nature and function of trophic relationships.

Measuring energy flow *(pages 31-32, 41-42)*

☐ 12. Understand the terms: **productivity**, **gross primary production**, **net primary production**, and **biomass**. Recognise how these relate to the transfer of energy to the next trophic level.

☐ 13. Explain how the energy flow in an ecosystem can be described quantitatively using an **energy flow diagram**. Include reference to:
  • Trophic levels (scaled boxes to illustrate relative amounts of energy at each level)
  • Direction of energy flow
  • Processes involved in energy transfer
  • Energy sources and energy sinks.

Ecological pyramids *(pages 43-44)*

☐ 14. Describe food chains quantitatively using **ecological pyramids**. Explain how these may be based on numbers of organisms, biomass of organisms, or energy content of organisms at each trophic level. Identify problems with the use of number pyramids, and understand why pyramids of biomass or energy are usually preferable.

☐ 15. Given appropriate information, construct and interpret ecological pyramids for different communities. Include:
  (a) Energy pyramids
  (b) Pyramids of numbers
  (c) Biomass pyramids
  Express the energy available at each trophic level in appropriate units. Identify the relationship between each of these types of pyramids and their corresponding food chains and webs. Explain why the shape of each graph is a pyramid (or sometimes an inverse pyramid).

Biogeochemical Cycles

☐ 16. Explain the terms **nutrient cycle** and **environmental reservoir**. Draw and interpret a generalised model of a nutrient cycle, identifying the roles of **primary productivity** and **decomposition** in nutrient cycling.

☐ 17. Using named examples, describe the general role of **saprotrophs** and **detritivores** in nutrient cycling.

The carbon cycle *(pages 45-46)*

☐ 18. Using a diagram, describe the stages in the **carbon cycle**, identifying the form of carbon at the different stages, and using arrows to show the direction of nutrient flow and labels to identify the processes involved. Identify the role of microorganisms, carbon sinks, and carbonates in the cycle.

☐ 19. Identify factors influencing the rate of carbon cycling. Recognise the role of respiration and photosynthesis in the short-term fluctuations and in the long-term global balance of oxygen and carbon dioxide.

☐ 20. Explain how humans may intervene in the carbon cycle and describe the effects of these interventions.

The nitrogen cycle (pages 47-48)

☐ 21. Using a diagram, describe the stages in the **nitrogen cycle**, identifying the form of nitrogen at the different stages, and using arrows to show the direction of nutrient flow and labels to identify the processes involved. Identify and explain the role of microorganisms in the cycle, as illustrated by:

(a) **Nitrifying bacteria** (*Nitrosomonas, Nitrobacter*)

(b) **Nitrogen-fixing bacteria** (*Rhizobium, Azotobacter*)

(c) **Denitrifying bacteria** (*Pseudomonas, Thiobacillus*).

☐ 22. Explain how humans may intervene in the nitrogen cycle and describe the effects of these interventions.

The phosphorus cycle (page 49)

☐ 23. Using diagrams, describe the **phosphorus cycle**, using arrows to show the direction of nutrient flow and labels to identify the processes involved. Identify the role of microorganisms in the cycle and contrast the phosphorus cycle with other nutrient cycles.

☐ 24. Explain how humans may intervene in the phosphorus cycle and describe the effects of these interventions.

The water cycle (page 50)

☐ 25. Describe the features of the **water** (hydrologic) **cycle**. Understand the ways in which water is cycled between various reservoirs and describe the major processes involved, including **evaporation**, **condensation**, **precipitation**, and **runoff**.

☐ 26. Explain how humans may intervene in the water cycle and describe the effects of these interventions.

See page 7 for additional details of these texts:

■ Allen, D, M. Jones, and G. Williams, 2001. **Applied Ecology**, chpt. 4.

■ Reiss, M. & J. Chapman, 2000. **Environmental Biology** (Cambridge Uni. Press), pp. 28-30, 36-37.

■ Smith, R. L. & T.M. Smith, 2001. **Ecology and Field Biology**, reading as required.

See page 7 for details of publishers of periodicals:

STUDENT'S REFERENCE

■ **Ecosystems** Biol. Sci. Rev., 9(4) March 1997, pp. 9-14. *Ecosystem structure including food chains and webs, nutrient cycles and energy flows, and ecological pyramids.*

■ **The Other Side of Eden** Biol. Sci. Rev., 15(3) February 2003, pp. 2-7. *An account of the Eden Project – the collection of artificial ecosystems in Cornwall. Its aims, future directions, and its role in the study of natural ecosystems and ecosystem modeling are discussed.*

■ **Ultimate interface** New Scientist, 14 November 1998 (Inside Science). *An excellent account of the cycling of chemicals in the biosphere, with particular emphasis on the role of the soil in nutrient cycling processes.*

■ **The Carbon Cycle** New Scientist, 2 Nov. 1991 (Inside Science). *Carbon is a vital element; its role in the world's ecosystems is explored in this four page, easy-to-read article.*

■ **A Users Guide to the Carbon Cycle** Biol. Sci. Rev., 10(1) Sept. 1997, pp. 10-13. *The carbon cycle: fixation and return to the atmosphere. The intervention of humans in the cycle is discussed.*

Presentation MEDIA
to support this topic:

ECOLOGY
• Communities

■ **The Case of the Missing Carbon** National Geographic, 205(2), Feb. 2004, pp. 88-117. *8 billion tonnes of CO_2 is dumped into the atmosphere annually, but only 3.5 billion tonnes remains there. The rest is absorbed into the Earth's carbon sinks.*

■ **Dung Beetles: Nature's Recyclers** Biol. Sci. Rev., 14(4) April 2002, pp. 31-33. *The ecological role of dung beetles in the recycling of nutrients.*

■ **Microorganisms in Agriculture** Biol. Sci. Rev., 11(5) May 1999, pp. 2-4. *The vital role of microorganisms in the ecology of communities is discussed (includes a discussion of N-fixation and the role of microorganisms in this process).*

■ **Nitrates in Soil and Water** New Scientist, 15 Sept. 1990 (Inside Science). *Nitrates and their role in the nitrogen cycle.*

■ **The Nitrogen Cycle** Biol. Sci. Rev., 13(2) November 2000, pp. 25-27. *An excellent account of the nitrogen cycle: conversions, role in ecosystems, and the influence of human activity.*

■ **Fish Predation** Biol. Sci. Rev., 14(1) September 2001, pp. 10-14. *Some fish species in freshwater systems in the UK are important top predators and can heavily influence the dynamics of the entire ecosystem.*

TEACHER'S REFERENCE

■ **One Rate to Rule Them All** New Scientist, 1 May 2004, pp. 38-41. *A universal law governing metabolic rates helps to explain energy flow through ecosystems now and in the future.*

■ **Whale of an Appetite** New Scientist, 24 October 1998, p. 25. *A crash in seal populations in the Alaskan region has had serious ramifications to food web stability: the orcas have prey switched from seals to sea otters, the otters have declined, and the sea urchin populations have exploded.*

■ **Kingdom of the Krill** New Scientist, 17 April 1999, pp. 36-41. *The details of the role of this marine organism in ocean food chains.*

■ **Waste Not** New Scientist, 29 August 1998, pp. 26-30. *Human sewage is often spread on land as fertilizer, but this carries the risk of passing pathogens through the food chain.*

■ **Ultimate Sacrifice** New Scientist, 6 September 1997, pp. 39-41. *The Alaskan salmon plays an important part in fuelling lake food webs. These salmon pass on vital nutrients obtained from the sea back into the food chain of the high latitude freshwater environments.*

■ **Terraria and Aquaria as Models for Teaching Relationships between Ecosystem Structure and Function** The American Biology Teacher, 59(1), Jan. 1997, pp. 52-53. *Using classroom set-ups as models for examining ecosystem function: photosynthesis, respiration, and nutrient cycling.*

■ **Global Population and the Nitrogen Cycle** Scientific American, July 1997, pp. 58-63. *The over use of nitrogen fertilizers is disturbing the nutrient balance in ecosystems and leading to pollution.*

■ **A Demonstration of Nitrogen Dynamics in Oxic and Hypoxic Soils and Sediments** The American Biology Teacher, 63(3), March, 2001, pp. 199-206. *A practical scheme aimed at improving student understanding of the nitrogen cycle.*

■ **The Ocean's Invisible Forest** Scientific American, August 2002, pp. 38-45. *The role of marine plankton in the global carbon cycle: ocean productivity and its influence on global climate.*

■ **Capturing Greenhouse Gases** Scientific American, February 2000, pp. 54-61. *New approaches to human interference in the carbon cycle: storing excess gases underground or in the oceans. Includes a useful synopsis of carbon transfers and how humans could intervene to alleviate the build-up of greenhouse gases.*

See pages 4-5 for details of how to access **Bio Links** from our web site: **www.thebiozone.com**. From Bio Links, access sites under the topics:

GENERAL BIOLOGY ONLINE RESOURCES > Online Textbooks and Lecture Notes: • An on-line biology book • Biology online.org • Kimball's biology pages • Mr Biology's biology web site • S-cool! A level biology revision guide ... *and others* **> Glossaries:** • Ecology glossary

ECOLOGY: • Ken's bioweb referencing • NatureWorks > **Energy Flows and Nutrient Cycles:** • A marine food web • Bioaccumulation • Human alteration of the global nitrogen cycle • Nitrogen: The essential element • The carbon cycle • The nitrogen cycle • The water cycle • Trophic pyramids and food webs ... *and others*

For the topic *The Dynamics of Populations,* **see the following sites: ECOLOGY >** **Populations and Communities:** • Bull Shoals Lake 1995 report • Communities • Competition • Death squared • Interactions • Intraspecific relations: Cooperation and competition • Population ecology • Quantitative population ecology • Species interactions • Death squared

Energy Flow and Nutrient Cycles

Energy in Ecosystems

An ecosystem is a natural unit of living (biotic) components, together with all the non-living (abiotic) components with which they interact. Two processes central to ecosystem function are **energy flow** and **chemical cycling**. The mitochondria of eukaryotic cells use the organic products of photosynthesis as fuel for cellular respiration. Respiration generates ATP; an energy currency for cellular work. Cellular work generates heat which is lost from the system. The waste products of cellular respiration are used as the raw materials for photosynthesis (see diagram below). Chemical elements such as nitrogen, phosphorus, and carbon are cycled between the biotic and abiotic components of the ecosystem. Energy, unlike matter, cannot be recycled. Ecosystems must receive a constant input of new energy from an outside source. In most cases, this is the sun.

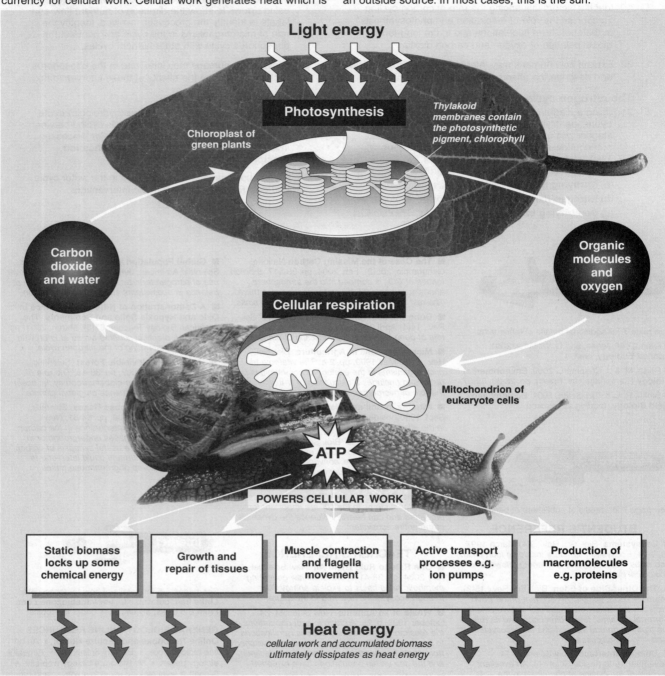

Light energy

Photosynthesis

Chloroplast of green plants

Thylakoid membranes contain the photosynthetic pigment, chlorophyll

Carbon dioxide and water

Organic molecules and oxygen

Cellular respiration

Mitochondrion of eukaryote cells

ATP

POWERS CELLULAR WORK

| Static biomass locks up some chemical energy | Growth and repair of tissues | Muscle contraction and flagella movement | Active transport processes e.g. ion pumps | Production of macromolecules e.g. proteins |

Heat energy
cellular work and accumulated biomass ultimately dissipates as heat energy

1. Write a word equation for each of the following processes. Include the energy source, raw materials, and waste products:

   (a) Photosynthesis: _____

   (b) Respiration: _____

2. Explain why ecosystems require a constant input of energy from an external source: _____

   _____

3. With respect to cycling of matter, explain the relationship between photosynthesis and respiration: _____

   _____

Code: A 2

Plant Productivity

The energy entering ecosystems is fixed by producers in photosynthesis. The rate of photosynthesis is dependent on factors such as temperature and the amount of light, water, and nutrients. The total energy fixed by a plant through photosynthesis is referred to as the **gross primary production (GPP)** and is usually expressed as Jm^{-2} (or kJm^{-2}), or as gm^{-2}. However, a portion of this energy is required by the plant for respiration. Subtracting respiration from GPP gives the **net primary production** (NPP). The **rate** of biomass production, or **net primary productivity**, is the biomass produced per area per unit time.

Measuring Productivity

Primary productivity of an ecosystem depends on a number of interrelated factors (light intensity, nutrients, temperature, water, and mineral supplies), making its calculation extremely difficult. Globally, the least productive ecosystems are those that are limited by heat energy and water. The most productive ecosystems are systems with high temperatures, plenty of water, and non-limiting supplies of soil nitrogen. The primary productivity of oceans is lower than that of terrestrial ecosystems because the water reflects (or absorbs) much of the light energy before it reaches and is utilised by producers. The table below compares the difference in the net primary productivity of various ecosystems.

Ecosystem Type	Net Primary Productivity	
	kcal m^{-2} y^{-1}	kJ m^{-2} y^{-1}
Tropical rainforest	15 000	63 000
Swamps and marshes	12 000	50 400
Estuaries	9000	37 800
Savanna	3000	12 600
Temperate forest	6000	25 200
Boreal forest	3500	14 700
Temperate grassland	2000	8400
Tundra/cold desert	500	2100
Coastal marine	2500	10 500
Open ocean	800	3360
Desert	< 200	< 840

* Data compiled from a variety of sources.

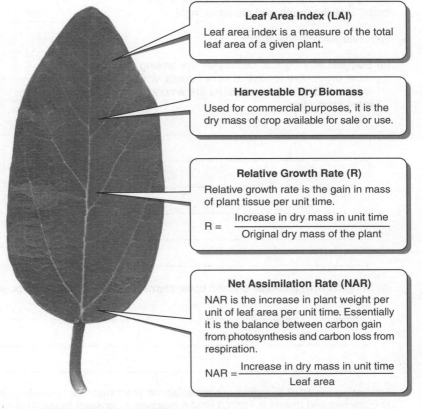

Leaf Area Index (LAI)
Leaf area index is a measure of the total leaf area of a given plant.

Harvestable Dry Biomass
Used for commercial purposes, it is the dry mass of crop available for sale or use.

Relative Growth Rate (R)
Relative growth rate is the gain in mass of plant tissue per unit time.
$$R = \frac{\text{Increase in dry mass in unit time}}{\text{Original dry mass of the plant}}$$

Net Assimilation Rate (NAR)
NAR is the increase in plant weight per unit of leaf area per unit time. Essentially it is the balance between carbon gain from photosynthesis and carbon loss from respiration.
$$NAR = \frac{\text{Increase in dry mass in unit time}}{\text{Leaf area}}$$

Net Primary Productivity of Selected Ecosystems (figures are in kJ m^{-2} y^{-1})

< 2500	< 12 500 – 42 000	< 42 000 – 105 000	2500 – 42 000
Arid desert	Temperate forest	Tropical rain forest	Continental shelf waters
Polar tundra and ice desert	Grassland agriculture	Intensive horticulture	Open ocean

1. Briefly describe three factors that may affect the primary productivity of an ecosystem:

   (a) _____

   (b) _____

   (c) _____

2. Explain the difference between **productivity** and **production** in relation to plants: _____

   _____

   _____

Code: DA 2

3. Suggest how the LAI might influence the rate of primary production: _____

_____

4. Using the data table on the previous page, choose a suitable graph format and plot the differences in the net primary productivity of various ecosystems (use either of the data columns provided, but not both). Use the graph grid provided, right.

5. With reference to the graph:

   (a) Suggest why tropical rainforests are among the most productive terrestrial ecosystems, while tundra and desert ecosystems are among the least productive:

   _____

   _____

   _____

   _____

   _____

   _____

   (b) Suggest why, amongst aquatic ecosystems, the NPP of the open ocean is low relative to that of coastal systems:

   _____

   _____

   _____

6. Estimating the NPP is relatively simple: all the plant material (including root material) from a measured area (e.g. 1 m^2) is collected and dried (at 105°C) until it reaches a constant mass. This mass, called the **standing crop**, is recorded (in kg m^{-2}). The procedure is repeated after some set time period (e.g. 1 month). The difference between the two calculated masses represents the *estimated* NPP:

   (a) Explain why the plant material was dried before weighing: _____

   _____

   (b) Define the term **standing crop**: _____

   (c) Suggest why this procedure only provides an estimate of NPP: _____

   _____

   (d) State what extra information would be required in order to express the standing crop value in kJ m^{-2}: _____

   _____

   (e) Suggest what information would be required in order to calculate the GPP: _____

   _____

7. Intensive horticultural systems achieve very high rates of production (about 10X those of subsistence systems).

   (a) Outline the means by which these high rates are achieved: _____

   _____

   (b) Comment on the sustainability of these high rates (summary of a group discussion if you wish): _____

   _____

   _____

   _____

Food Chains

Every ecosystem has a **trophic structure**: a hierarchy of feeding relationships which determines the pathways for energy flow and nutrient cycling. Species are divided into trophic levels on the basis of their sources of nutrition. The first trophic level (**producers**), ultimately supports all other levels. The consumers are those that rely on producers for their energy. Consumers are ranked according to the trophic level they occupy (first order, second order, etc.). The sequence of organisms, each of which is a source of food for the next, is called a **food chain**. Food chains commonly have four links but seldom more than six. Those organisms whose food is obtained through the same number of links belong to the same trophic level. Note that some consumers (particularly "top" carnivores and omnivores) may feed at several different trophic levels, and many primary consumers eat many plant species. The different food chains in an ecosystem therefore tend to form complex webs of interactions (food webs).

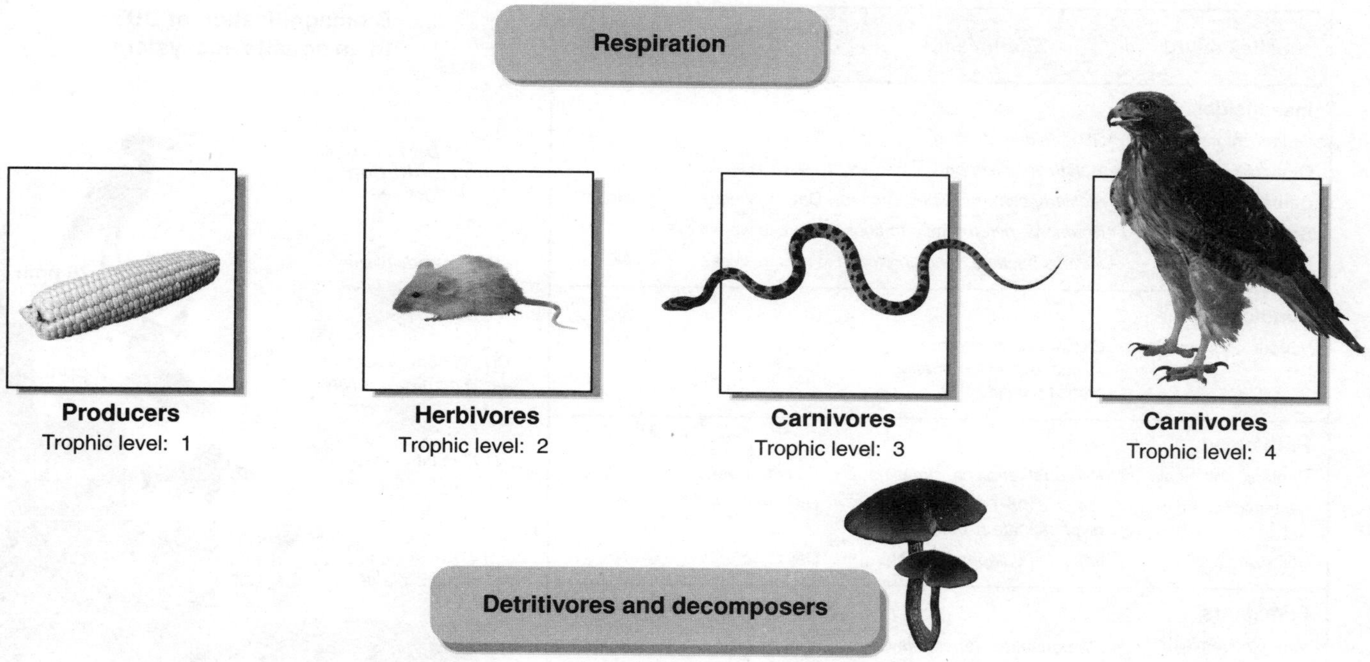

Respiration

Producers
Trophic level: 1

Herbivores
Trophic level: 2

Carnivores
Trophic level: 3

Carnivores
Trophic level: 4

Detritivores and decomposers

The diagram above represents the basic elements of a food chain. In the questions below, you are asked to add to the diagram the features that indicate the flow of energy through the community of organisms.

1. (a) State the original energy source for this food chain: _____

   (b) Draw arrows on the diagram above to show how the energy flows through the organisms in the food chain.
   (c) Label each of the arrows with the process that carries out this transfer of energy.
   (d) Draw arrows on the diagram to show how the energy is lost by way of respiration.

2. (a) Describe what happens to the **amount** of energy available to each successive trophic level in a food chain:

   _____

   (b) Explain why this is the case: _____

   _____

3. Discuss the trophic structure of ecosystems, including reference to **food chains** and **trophic** levels:

   _____
   _____
   _____
   _____
   _____
   _____
   _____

4. If the eagle (above) was found to eat both snakes and mice, explain what you could infer about the tropic level(s) it occupied:

   _____

Energy Flow and Nutrient Cycles

Code: A 2

Pesticides and Bioaccumulation

Certain substances in the environment are harmful when absorbed in high concentrations. Substances, such as pesticides, radioactive isotopes, heavy metals, and industrial chemicals such as PCBs can be taken up by organisms via their food or simply absorbed from the surrounding medium. The **toxicity** of a pesticide is a measure of how poisonous the chemical is, not only to the target organisms, but to non-target species as well. The **specificity** (broad or narrow spectrum) of a pesticide describes how selective it is in targeting a pest. An important issue relating to the use of a pesticide is its **persistence** (how long it stays in the environment). A pesticide may be *biodegradable* or resistant to biological breakdown. Many highly persistent pesticides cannot be metabolised or excreted. Instead, they are stored in fatty tissues and have the potential for **bioaccumulation** (biological magnification); a progressive concentration with increasing trophic level. Higher order consumers (e.g. predatory birds and mammals) may ingest harmful or lethal quantities of a chemical because they eat a large number of lower order consumers.

Pesticide type	Examples	Environmental persistence	Biomag-nification
Insecticides			
Organochlorines	DDT*, aldrin, dieldrin	2-15 yrs	Yes
Organophosphates	Malathion, diazinon	1-2 weeks/years	No
Carbamates	Aldicarb, carbaryl, zineb, maneb	Days to weeks	No
Botanicals	Rotenone, pyrethrum, camphor	Days to weeks	No
Microbials	Various bacteria, fungi, protozoa	Days to weeks	No
Fungicides			
Various chemicals	Captan, zeneb, carbon sulfide, pentachlorphenol, methyl bromide	Days	No
Herbicides			
Contact§ chemicals	Paraquat, atrazine, simazine	Days to weeks	No
Systemic¶ chemicals	2,4-D, 2,4,5-T, Silvex, diruon, glyphosphate (Roundup)	Days to weeks	No
Soil sterilants	Tribualin, dalapon, butylate	Days	No
Fumigants			
Various chemicals	Carbon tetrachloride, ethylene dibromide, methyl bromide	Years	Yes

* Now banned in most developed countries
¶ Systemic chemicals: Effective when it enters the general circulation of the plant or animal
§ Contact chemicals: Effective when it comes in contact with surface tissue

Source of Table Data: *Living in the Environment* (11th Ed.), G. Tyler Miller, Jr., Brooks/Cole Publishing (2000)

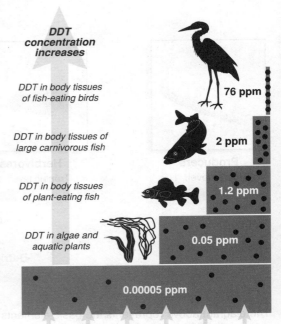

Biomagnification of DDT in an aquatic ecosystem

DDT concentration increases

DDT in body tissues of fish-eating birds — 76 ppm

DDT in body tissues of large carnivorous fish — 2 ppm

DDT in body tissues of plant-eating fish — 1.2 ppm

DDT in algae and aquatic plants — 0.05 ppm

0.00005 ppm

DDT enters the lake as runoff from farmland which has been sprayed with the insecticide

1. Define the following terms as they relate to the characteristics and use of pesticides:

   (a) Toxicity: _____

   (b) Specificity: _____

   (c) Biodegradable: _____

   (d) Bioaccumulation: _____

   (e) Contact chemical: _____

   (f) Systemic chemical: _____

2. Calculate the increase in DDT concentration between each step in the food chain:

   (a) Between water and algae: _____ *0.05 ÷ 0.00005 = 1000 times* _____

   (b) Between algae and plant-eating fish: _____

   (c) Between plant-eating fish and carnivorous fish: _____

   (d) Between carnivorous fish and fish-eating birds: _____

3. Suggest why some **organochlorine** pesticides have been banned in most developed countries: _____

   _____

4. Explain why top consumers are most at risk from bioaccumulation: _____

   _____

Constructing a Food Web

The actual species inhabiting any particular lake may vary depending on locality, but certain types of organisms (as shown below) are typically represented. For the sake of simplicity, only fifteen organisms are represented here. Real lake communities may have hundreds of different species interacting together. The bulk of this species diversity is in the lower trophic levels (producers and invertebrate grazers). Your task is to assemble the organisms below into a food web, in a way that illustrates

their trophic status and their relative trophic position(s). The food resource represented by **detritus** is not shown here. Detritus comprises the accumulated debris of dead organisms in varying stages of decay. This debris may arise from within the lake itself or it may be washed in from the surrounding lake margins and streams. The detritus settles through the water column and eventually forms a layer on the lake bottom. It provides a rich food source for any organism that can exploit it.

Feeding Requirements of Lake Organisms

Daphnia
Small freshwater crustacean that forms part of the zooplankton. It feeds on planktonic algae by filtering them from the water with its limbs.

Autotrophic protoctists
e.g. Chlamydomonas, Euglena (pictured)
Two of genera that form the phytoplankton (commonly called algae or "plant plankton").

Asplanchna (planktonic rotifer)
A large, carnivorous rotifer that feeds on protozoa and young zooplankton (e.g. *Daphnia*). Note that most rotifers are small herbivores.

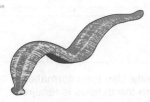

Leech (*Glossiphonia*)
Leeches are fluid feeding predators of smaller invertebrates, including rotifers, small pond snails and worms.

Macrophytes (various species)
A variety of flowering aquatic plants are adapted for being submerged, free-floating, or growing at the lake margin.

Three-spined stickleback (*Gasterosteus*)
A common fish of freshwater ponds and lakes. It feeds mainly on small invertebrates such as *Daphnia* and insect larvae.

Diving beetle (*Dytiscus*)
Diving beetles feed on aquatic insect larvae and adult insects blown into the lake community. The will also eat organic detritus collected from the bottom mud.

Carp (*Cyprinus*)
A heavy bodied freshwater fish that feeds mainly on bottom living insect larvae and snails, but will also take some plant material (not algae).

Dragonfly larva
Large aquatic insect larvae that are voracious predators of small invertebrates including *Hydra*, *Daphnia*, other insect larvae, and leeches.

Great pond snail (*Limnaea*)
Omnivorous pond snail, eating both plant and animal material, living or dead, although the main diet is aquatic macrophytes.

Herbivorous water beetles (e.g. *Hydrophilus*)
Feed on water plants, although the young beetle larvae are carnivorous, feeding primarily on small pond snails.

Protozan (e.g. *Paramecium*)
Ciliated protozoa such as *Paramecium* feed primarily on bacteria and microscopic green algae such as *Chlamydomonas*.

Pike (*Esox lucius*)
A top ambush predator of all smaller fish and amphibians, although they are also opportunistic predators of rodents and small birds.

Mosquito larva
(*Culex* spp.)
The larvae of most mosquito species, e.g. *Culex*, feed on planktonic algae before passing through a pupal stage and undergoing metamorphosis into adult mosquitoes.

Hydra
A small carnivorous cnidarian that captures small prey items such as small *Daphnia* and insect larvae using its stinging cells on the tentacles.

Energy Flow and Nutrient Cycles

Code: A 2

Instructions

1. (a) Read the information provided for each species on the previous page, taking note of what it feeds on.

    (b) Identify the **producer** species present, as well as herbivores, carnivores, and omnivores.

    (c) Starting with **producer** species, construct **4** different **food chains** (using their names only) to show the feeding relationships between the organisms (NOTE: some food chains may be shorter than others; some species will be repeated in one or more subsequent food chains). An example of a food chain has already been completed for you.

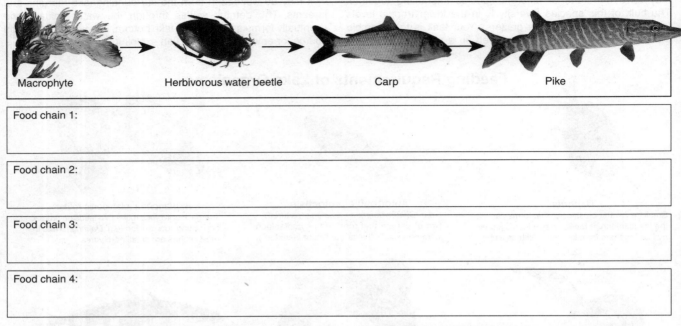

Macrophyte → Herbivorous water beetle → Carp → Pike

Food chain 1:

Food chain 2:

Food chain 3:

Food chain 4:

2. (a) Use the food chains created above to help you to draw up a **food web** for this community. Use the information supplied to draw arrows showing the flow of **energy** between species (only energy **from** the detritus is required).

    (b) Label each species to indicate its position in the food web, i.e. its trophic level (**T1, T2, T3, T4, T5**). Where a species occupies more than one trophic level, indicate this, e.g. **T2/3**:

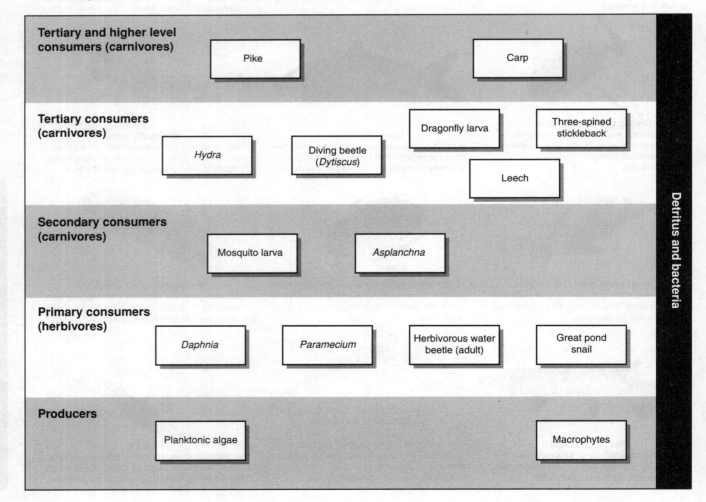

Tertiary and higher level consumers (carnivores)
Pike   Carp

Tertiary consumers (carnivores)
Hydra   Diving beetle (*Dytiscus*)   Dragonfly larva   Three-spined stickleback   Leech

Secondary consumers (carnivores)
Mosquito larva   *Asplanchna*

Primary consumers (herbivores)
Daphnia   *Paramecium*   Herbivorous water beetle (adult)   Great pond snail

Producers
Planktonic algae   Macrophytes

Detritus and bacteria

Dingo Food Webs

Dingoes are widespread in Australia and occupy the position as top predator in their food web. In different parts of Australia, dingoes are part of varied communities, each comprising a different food web from the other. Below are samples taken from six different locations to illustrate how the dingo's diet, and therefore the food web of which it is a part, varies from one location to another. In some cases, not all prey species are listed (e.g. arid & semi-arid central Australia had another 62 species).

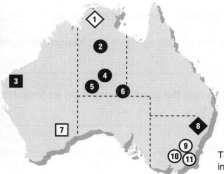

Location of sampling sites
1 Kakadu National Park
2 Barkley Tableland
3 Fortesque River
4 Harts Ranges
5 Eldunda
6 Simpson Desert
7 Nullarbor Plain
8 George's Creek Nature Reserve
9 Kosciusko National Park
10 Victorian Highlands
11 Nadgee Nature Reserve

Despite the immense range of potential prey species across Australia, only ten species formed almost 80% of a dingo's diet. Dingoes are specialists, rather than generalists, with respect to dietary intake. However, they use a large range of hunting tactics to catch prey and are considered to be generalists in this respect.

The table below shows the diet of dingoes in six major Australian habitats. The numbers represent the proportion (**% occurrence**) of stomachs or faeces that contained each prey species. A total of 12 802 stomachs and faeces were sampled over a 20 year period (1966-1986).

◇ Wet-dry tropics North Australia	6722 faeces	● Arid & semi-arid central Australia	1480 stomachs	☐ Arid south-west Australia	131 faeces/ stomachs
Dusky rat	33.8	Rabbit	37.9	Rabbit	63.4
Magpie goose	32.5	Cattle	23.3	Red kangaroo	32.1
Agile wallaby	15.1	Long-haired rat	17.6	Cattle	7.6
Northern ringtailed possum	9.7	Red kangaroo	10.2	Red fox	3.8
Grass species	7.1	Central netted dragon	7.8	Little crow	2.3
Feral water buffalo	5.8	Small mammal (undetermined sp.)	3.8	Bobtail skink	1.5
Feral pig	3.5	House mouse	3.6	Feral cat	1.5
Unidentified matter	2.6	Grasshopper	2.7	Centipede/millipede	0.8
Antilopine wallaroo	1.8	Bearded dragon	2.2	Dingo	0.8
Northern brown bandicoot	1.4	Zebra finch	2.1	Grasshopper	0.8
Feral cattle	1.3	Bird (undetermined species)	2.0		
Bird (undetermined species)	0.7	Feral cat	1.8		
Insect (undetermined species)	0.7	Galah	1.8		
Beetle	0.4	Budgerigar	1.5		

■ Semi-arid north-west Australia	413 faeces/ stomachs	○ Cool coastal mountains SE Australia	2063 faeces/ stomachs	◆ Humid coastal mountains eastern Australia	1993 faeces
Red kangaroo/Euro	80.6	Swamp wallaby	17.9	Swamp wallaby	30.5
Cattle	11.4	Wallaby (undetermined spp.)	15.8	Bush rat	12.2
Sheep	8.0	Wombat	15.0	Red-necked wallaby	11.1
Bird (undetermined species)	5.6	Animal remains (unidentified)	11.5	Brushtail possums	6.9
Reptile (undetermined species)	3.4	Rabbit	10.5	Bandicoots (long-nosed, Southern brown)	6.8
Insects (undetermined species)	2.9	Common ringtail possum	8.0	Rabbit	6.4
Echidna	2.2	Waterbird (undetermined sp.)	7.7	Antichinuses (brown, dusky)	5.8
Dingo	1.7	Red-neck wallaby	5.3	Parma wallaby	4.6
Feral cat	0.5	Possum (undetermined species)	5.1	Common ringtail possum	4.4
Bat (undetermined species)	0.2	Rat (undetermined species)	5.0	Ring-necked pademelon	3.8
Fish (undetermined species)	0.2	Little penguin	4.4	Echidna	3.5
Red fox	0.2	Fish (undetermined species)	3.7	Long-nosed potoroo	1.7
Rothschild's rock wallaby	0.2	Mutton bird	3.6	Greater glider	1.5
		Echidna	3.3		

Adapted from: Corbett, L. 1995. *The dingo in Australia and Asia*, Appendix C: pp. 183-186. University of NSW Press.

1. Sites 7 and 3 (above) yielded a small number of prey species in the samples taken. Suggest the likely reason for this:

_____

2. Name **three** prey species that are taken by dingoes in the '*Cool coastal mountains, SE Australia*' (sites 9, 10, and 11), that are restricted to that type of environment (i.e. not represented in the prey taken at other sites):

_____

Energy Flow and Nutrient Cycles

Code: RDA 2

3. Explain what evidence there is from the data on the previous page, that dingoes engage in cannibalism:

_____

4. State what general kind of prey makes up most of the dingoes' diet: _____

5. At some sample sites, the dingoes' prey included domesticated animals.

   (a) Name the prey species that represent domesticated animals: _____

   (b) State what kind of environmental conditions have encouraged dingoes to make these animals part of their diet:

   _____

6. (a) State which of the sample sites has the least reliable data for indicating the diet of dingoes in its area:

   _____

   (b) Explain why you made your choice: _____

   _____

7. In this study, the diet of dingoes was determined by the sampling methods of examining large numbers of stomach contents and faeces.

   (a) Explain which of these two methods should prove the most reliable for positive identification of prey species:

   _____

   (b) Suggest **two reasons** why the researchers did not simply follow the dingoes and watch what they ate as a way of gathering dietary information on the dingoes:

   Reason 1: _____

   Reason 2: _____

8. Using the data on the previous page, choose one of the 'regional ecotypes' (e.g. wet-dry tropics, north Australia) and produce a food web in the space below. Use only the first **five positively identified** prey species (in most cases, do not include unidentified species). This activity will require you to carry out some research into what the prey species eat.

Energy Inputs and Outputs

The way living things obtain their energy can be classified into two categories. The group upon which all others depend are called **producers** or **autotrophs**: organisms that are able to manufacture their food from simple inorganic substances. The **consumers** or **heterotrophs** (comprising the herbivores, carnivores, omnivores, decomposers, and detritivores), feed on the autotrophs or other heterotrophs to obtain their energy. The energy flow into and out of each trophic level in a food chain can be identified and represented diagrammatically using arrows of different sizes. The sizes of the arrows (see the diagrams below and on the next page) represent different amounts of energy lost from that particular trophic level.

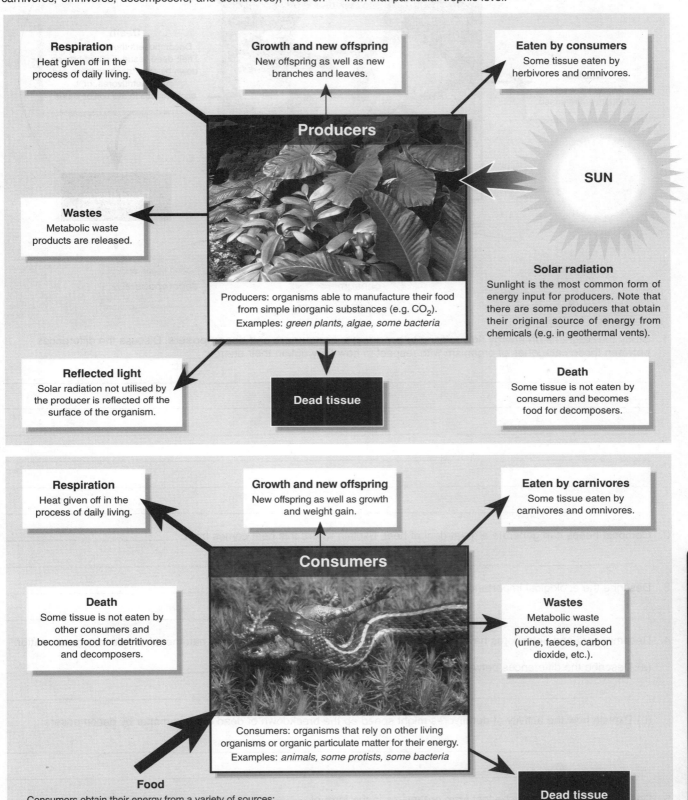

Respiration
Heat given off in the process of daily living.

Growth and new offspring
New offspring as well as new branches and leaves.

Eaten by consumers
Some tissue eaten by herbivores and omnivores.

Producers

SUN

Wastes
Metabolic waste products are released.

Solar radiation
Sunlight is the most common form of energy input for producers. Note that there are some producers that obtain their original source of energy from chemicals (e.g. in geothermal vents).

Producers: organisms able to manufacture their food from simple inorganic substances (e.g. CO_2).
Examples: *green plants, algae, some bacteria*

Reflected light
Solar radiation not utilised by the producer is reflected off the surface of the organism.

Dead tissue

Death
Some tissue is not eaten by consumers and becomes food for decomposers.

Respiration
Heat given off in the process of daily living.

Growth and new offspring
New offspring as well as growth and weight gain.

Eaten by carnivores
Some tissue eaten by carnivores and omnivores.

Consumers

Death
Some tissue is not eaten by other consumers and becomes food for detritivores and decomposers.

Wastes
Metabolic waste products are released (urine, faeces, carbon dioxide, etc.).

Consumers: organisms that rely on other living organisms or organic particulate matter for their energy.
Examples: *animals, some protists, some bacteria*

Food
Consumers obtain their energy from a variety of sources: plant tissues (**herbivores**), animal tissues (**carnivores**), plant and animal tissues (**omnivores**), dead organic matter or detritus (**detritivores** and **decomposers**).

Dead tissue

Energy Flow and Nutrient Cycles

Code: A 1

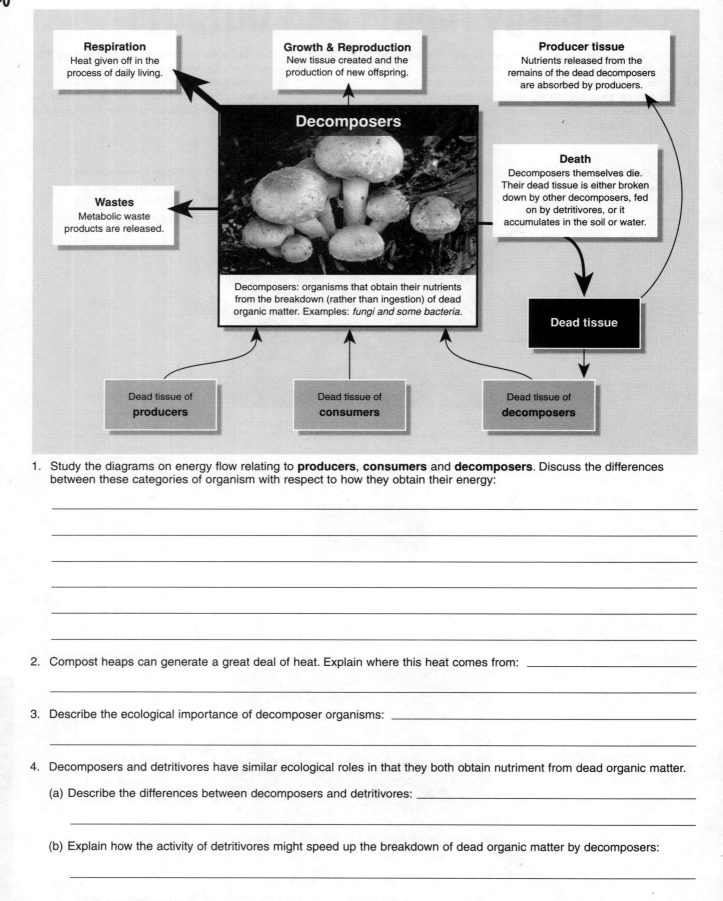

Respiration
Heat given off in the process of daily living.

Growth & Reproduction
New tissue created and the production of new offspring.

Producer tissue
Nutrients released from the remains of the dead decomposers are absorbed by producers.

Decomposers

Death
Decomposers themselves die. Their dead tissue is either broken down by other decomposers, fed on by detritivores, or it accumulates in the soil or water.

Wastes
Metabolic waste products are released.

Decomposers: organisms that obtain their nutrients from the breakdown (rather than ingestion) of dead organic matter. Examples: *fungi and some bacteria*.

Dead tissue

Dead tissue of **producers**

Dead tissue of **consumers**

Dead tissue of **decomposers**

1. Study the diagrams on energy flow relating to **producers**, **consumers** and **decomposers**. Discuss the differences between these categories of organism with respect to how they obtain their energy:

_____
_____
_____
_____
_____
_____

2. Compost heaps can generate a great deal of heat. Explain where this heat comes from: _____
_____

3. Describe the ecological importance of decomposer organisms: _____
_____

4. Decomposers and detritivores have similar ecological roles in that they both obtain nutriment from dead organic matter.

   (a) Describe the differences between decomposers and detritivores: _____
   _____

   (b) Explain how the activity of detritivores might speed up the breakdown of dead organic matter by decomposers:
   _____
   _____

5. Describe how energy may be lost from organisms in the form of:

   (a) Wastes: _____

   (b) Respiration: _____

Energy Flow in an Ecosystem

The flow of energy through an ecosystem can be measured and analysed. It provides some idea as to the energy trapped and passed on at each trophic level. Each trophic level in a food chain or web contains a certain amount of biomass: the dry weight of all organic matter contained in its organisms. Energy stored in biomass is transferred from one trophic level to another (by eating, defaecation etc.), with some being lost as low-grade heat energy to the environment in each transfer. Three definitions are useful:

- **Gross primary production**: The total of organic material produced by plants (including that lost to respiration).
- **Net primary production**: The amount of biomass that is available to consumers at subsequent trophic levels.

- **Secondary production**: The amount of biomass at higher trophic levels (consumer production). Production figures are sometimes expressed as rates (productivity).

The percentage of energy transferred from one trophic level to the next varies between 5% and 20% and is called the **ecological efficiency** (efficiency of energy transfer). An average figure of 10% is often used. The path of energy flow in an ecosystem depends on its characteristics. In a tropical forest ecosystem, most of the primary production enters the detrital and decomposer food chains. However, in an ocean ecosystem or an intensively grazed pasture more than half the primary production may enter the grazing food chain.

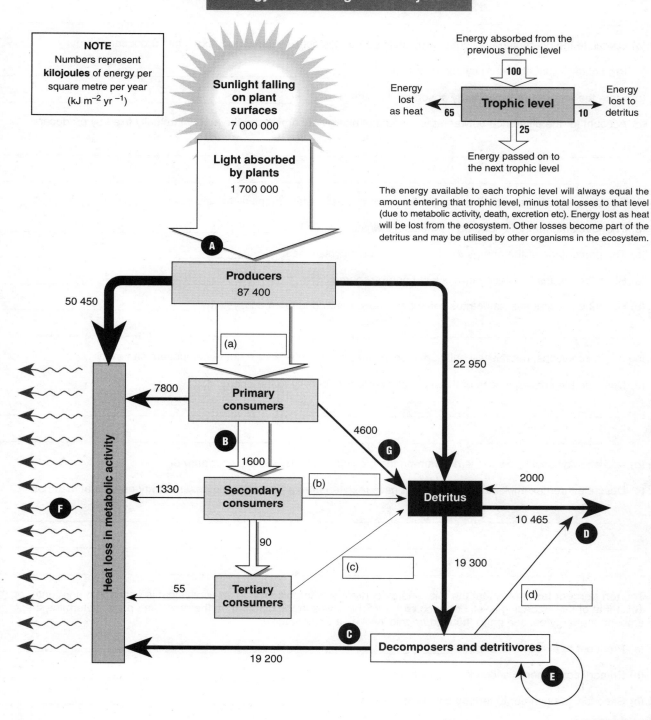

Energy Flow Through an Ecosystem

NOTE
Numbers represent **kilojoules** of energy per square metre per year ($kJ\ m^{-2}\ yr^{-1}$)

Energy absorbed from the previous trophic level
100
Energy lost as heat **65** | Trophic level | **10** Energy lost to detritus
25
Energy passed on to the next trophic level

The energy available to each trophic level will always equal the amount entering that trophic level, minus total losses to that level (due to metabolic activity, death, excretion etc). Energy lost as heat will be lost from the ecosystem. Other losses become part of the detritus and may be utilised by other organisms in the ecosystem.

Sunlight falling on plant surfaces 7 000 000

Light absorbed by plants 1 700 000

A

Producers 87 400

50 450

(a)

22 950

7800 → **Primary consumers**

B

1600

4600 → **G**

(b)

Detritus

2000

10 465 **D**

1330 → **Secondary consumers**

90

(c)

19 300

55 → **Tertiary consumers**

(d)

Heat loss in metabolic activity

F

C

Decomposers and detritivores

19 200

E

Energy Flow and Nutrient Cycles

Code: RDA 2

42

1. Study the diagram on the previous page illustrating energy flow through a hypothetical ecosystem. Use the example at the top of the page as a guide to calculate the missing values (a)–(d) in the diagram. Note that the sum of the energy inputs always equals the sum of the energy outputs. Place your answers in the spaces provided on the diagram.

2. Describe the original source of energy that powers this ecosystem: _____

3. Identify the processes that are occurring at the points labelled **A – G** on the diagram:

   A. _____   E. _____

   B. _____   F. _____

   C. _____   G. _____

   D. _____

4. (a) Calculate the percentage of light energy falling on the plants that is absorbed at point **A**:

   Light absorbed by plants ÷ sunlight falling on plant surfaces x 100 = _____

   (b) Describe what happens to the light energy that is not absorbed: _____

   _____

5. (a) Calculate the percentage of light energy absorbed that is actually converted (fixed) into producer energy:

   Producers ÷ light absorbed by plants x 100 = _____

   (b) State the **amount** of light energy absorbed that is **not** fixed: _____

   (c) Account for the difference between the amount of energy absorbed and the amount actually fixed by producers:

   _____

   _____

6. Of the total amount of energy **fixed** by producers in this ecosystem (at point **A**) calculate:

   (a) The total amount that ended up as metabolic waste heat (in kJ): _____

   (b) The percentage of the energy fixed that ended up as waste heat: _____

7. (a) State the groups for which detritus is an energy source: _____

   (b) Describe by what means detritus could be removed or added to an ecosystem: _____

   _____

8. In certain conditions, detritus will build up in an environment where few (or no) decomposers can exist.

   (a) Describe the consequences of this lack of decomposer activity to the energy flow:

   _____

   _____

   (b) Add an additional arrow to the diagram on the previous page to illustrate your answer.

   (c) Describe three examples of materials that have resulted from a lack of decomposer activity on detrital material:

   _____

   _____

   _____

9. The **ten percent law** states that the total energy content of a trophic level in an ecosystem is only about one-tenth (or 10%) that of the preceding level. For each of the trophic levels in the diagram on the preceding page, determine the amount of energy passed on to the next trophic level as a percentage:

   (a) Producer to primary consumer: _____

   (b) Primary consumer to secondary consumer: _____

   (c) Secondary consumer to tertiary consumer: _____

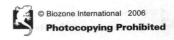

Ecological Pyramids

The trophic levels of any ecosystem can be arranged in pyramid of increasing trophic level. The first trophic level is placed at the bottom and subsequent trophic levels are stacked on top in their 'feeding sequence'. Ecological pyramids can illustrate changes in the numbers, biomass (weight), or energy content of organisms at each level. Each of these three kinds of pyramids tells us something different about the flow of energy and movement of materials between one trophic level and the next. The type of pyramid you choose in order to express information about an ecosystem will depend on what particular features of the ecosystem you are interested in and, of course, the type of data you have collected.

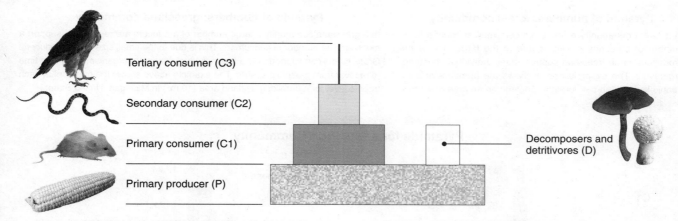

Tertiary consumer (C3)

Secondary consumer (C2)

Primary consumer (C1)

Primary producer (P)

Decomposers and detritivores (D)

The generalised ecological pyramid pictured above shows a conventional pyramid shape, with a large number (or biomass) of producers forming the base for an increasingly small number (or biomass) of consumers. Decomposers are placed at the level of the primary consumers and off to the side. They may obtain energy from many different trophic levels and so do not fit into the conventional pyramid structure. For any particular ecosystem at any one time (e.g. the forest ecosystem below), the shape of this typical pyramid can vary greatly depending on whether the trophic relationships are expressed as numbers, biomass or energy.

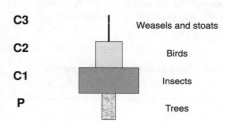

C3 — Weasels and stoats
C2 — Birds
C1 — Insects
P — Trees

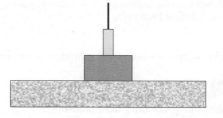

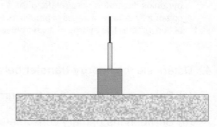

Numbers in a forest community

Pyramids of numbers display the number of individual organisms at each trophic level. The pyramid above has few producers, but they may be of a very large size (e.g. trees). This gives an 'inverted pyramid', although not all pyramids of numbers are like this.

Biomass in a forest community

Biomass pyramids measure the 'weight' of biological material at each trophic level. Water content of organisms varies, so 'dry weight' is often used. Organism size is taken into account, so meaningful comparisons of different trophic levels are possible.

Energy in a forest community

Pyramids of energy are often very similar to biomass pyramids. The energy content at each trophic level is generally comparable to the biomass (i.e. similar amounts of dry biomass tend to have about the same energy content).

1. Describe what the three types of ecological pyramids measure:

   (a) Number pyramid: _____

   (b) Biomass pyramid: _____

   (c) Energy pyramid: _____

2. Explain the advantage of using a biomass or energy pyramid rather than a pyramid of numbers to express the relationship between different trophic levels:

   _____

   _____

   _____

3. Explain why it is possible for the forest community (on the next page) to have very few producers supporting a large number of consumers:

   _____

   _____

Energy Flow and Nutrient Cycles

Code: DA 2

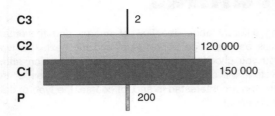

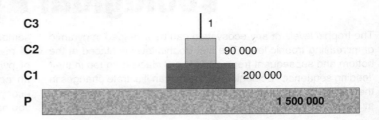

Pyramid of numbers: forest community

In a forest community a few producers may support a large number of consumers. This is due to the large size of the producers; large trees can support many individual consumer organisms. The example above shows the numbers at each trophic level for an oak forest in England, in an area of 10 m^2.

Pyramid of numbers: grassland community

In a grassland community a large number of producers are required to support a much smaller number of consumers. This is due to the small size of the producers. Grass plants can support only a few individual consumer organisms and take time to recover from grazing pressure. The example above shows the numbers at each trophic level for a derelict grassland area (10 m^2) in Michigan, United States.

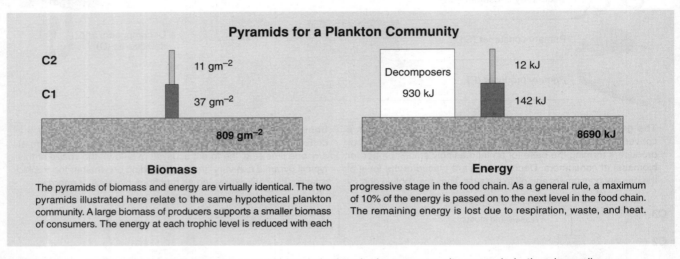

Pyramids for a Plankton Community

Biomass

Energy

The pyramids of biomass and energy are virtually identical. The two pyramids illustrated here relate to the same hypothetical plankton community. A large biomass of producers supports a smaller biomass of consumers. The energy at each trophic level is reduced with each progressive stage in the food chain. As a general rule, a maximum of 10% of the energy is passed on to the next level in the food chain. The remaining energy is lost due to respiration, waste, and heat.

4. Determine the **energy transfer** between trophic levels in the plankton community example in the above diagram:

(a) Between producers and the primary consumers: _____

(b) Between the primary consumers and the secondary consumers: _____

(c) Explain why the energy passed on from the producer to primary consumers is considerably less than the normally expected 10% occurring in most other communities (describe where the rest of the energy was lost to):

_____

_____

(d) After the producers, which trophic group has the greatest energy content: _____

(e) Give a likely explanation why this is the case: _____

_____

_____

An unusual biomass pyramid

The biomass pyramids of some ecosystems appear rather unusual with an inverted shape. The first trophic level has a lower biomass than the second level. What this pyramid does not show is the rate at which the producers (algae) are reproducing in order to support the larger biomass of consumers.

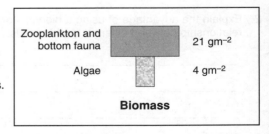

5. Give a possible explanation of how a small biomass of producers(algae) can support a larger biomass of consumers (zooplankton):

_____

_____

_____

The Carbon Cycle

Carbon is an essential element in living systems, providing the chemical framework to form the molecules that make up living organisms (e.g. proteins, carbohydrates, fats, and nucleic acids). Carbon also makes up approximately 0.03% of the atmosphere as the gas carbon dioxide (CO_2), and it is present in the ocean as carbonate and bicarbonate, and in rocks such as limestone. Carbon cycles between the living (biotic) and non-living (abiotic)

environment: it is fixed in the process of photosynthesis and returned to the atmosphere in respiration. Carbon may remain locked up in biotic or abiotic systems for long periods of time as, for example, in the wood of trees or in fossil fuels such as coal or oil. Human activity has disturbed the balance of the carbon cycle (the global carbon budget) through activities such as combustion (e.g. the burning of wood and **fossil fuels**) and deforestation.

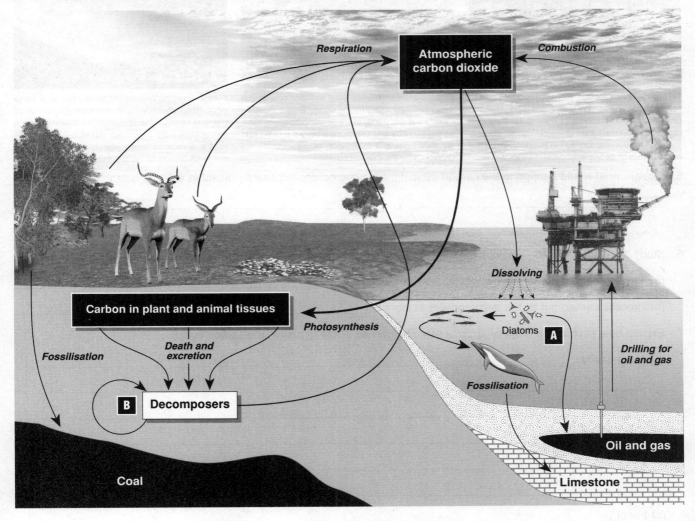

1. In the diagram above, add arrows and labels to show the following activities:

   (a) Dissolving of limestone by acid rain
   (b) Release of carbon from the marine food chain

   (c) Mining and burning of coal
   (d) Burning of plant material.

2. Describe the **biological origin** of the following geological deposits:

   (a) Coal: _____

   (b) Oil: _____

   (c) Limestone: _____

3. Describe the two processes that release carbon into the atmosphere: _____

   _____

   _____

4. Name the four geological reservoirs (sinks), in the diagram above, that can act as a source of carbon:

   (a) _____ (c) _____

   (b) _____ (d) _____

Energy Flow and Nutrient Cycles

Code: A 2

46

Termite mound in rainforest

Dung beetle on cow pat

Bracket fungus on tree trunk

Termites: These insects play an important role in nutrient recycling. With the aid of symbiotic protozoans and bacteria in their guts, they can digest the tough cellulose of woody tissues in trees. Termites fulfill a vital function in breaking down the endless rain of debris in tropical rainforests.

Dung beetles: Beetles play a major role in the decomposition of animal dung. Some beetles merely eat the dung, but true dung beetles, such as the scarabs and *Geotrupes*, bury the dung and lay their eggs in it to provide food for the beetle grubs during their development.

Fungi: Together with decomposing bacteria, fungi perform an important role in breaking down dead plant matter in the leaf litter of forests. Some mycorrhizal fungi have been found to link up to the root systems of trees where an exchange of nutrients occurs (a mutualistic relationship).

5. Explain what would happen to the carbon cycle if there were no decomposers present in an ecosystem:

6. Study the diagram on the previous page and identify the processes represented at the points labelled [**A**] and [**B**]:

   (a) Process carried out by the diatoms at label **A**:

   (b) Process carried out by the decomposers at label **B**:

7. Explain how each of the three organisms listed below has a role to play in the carbon cycle:

   (a) Dung beetles:

   (b) Termites:

   (c) Fungi:

8. In natural circumstances, accumulated reserves of carbon such as peat, coal and oil represent a **sink** or natural diversion from the cycle. Eventually the carbon in these sinks returns to the cycle through the action of geological processes which return deposits to the surface for oxidation.

   (a) Describe what effect human activity is having on the amount of carbon stored in sinks:

   (b) Explain two global effects arising from this activity:

   (c) Suggest what could be done to prevent or alleviate these effects:

The Nitrogen Cycle

Nitrogen is a crucial element for all living things, forming an essential part of the structure of proteins and nucleic acids. The Earth's atmosphere is about 80% nitrogen gas (N_2), but molecular nitrogen is so stable that it is only rarely available directly to organisms and is often in short supply in biological systems. Bacteria play an important role in transferring nitrogen between the biotic and abiotic environments. Some bacteria are able to fix atmospheric nitrogen, while others convert ammonia to nitrate and thus make it available for incorporation into plant and animal tissues. Nitrogen-fixing bacteria are found living freely in the soil *(Azotobacter)* and living symbiotically with some plants in root nodules *(Rhizobium)*. Lightning discharges also cause the oxidation of nitrogen gas to nitrate which ends up in the soil. Denitrifying bacteria reverse this activity and return fixed nitrogen to the atmosphere. Humans intervene in the nitrogen cycle by producing, and applying to the land, large amounts of nitrogen fertiliser. Some applied fertiliser is from organic sources (e.g. green crops and manures) but much is inorganic, produced from atmospheric nitrogen using an energy-expensive industrial process. Overuse of nitrogen fertilisers may lead to pollution of water supplies, particularly where land clearance increases the amount of leaching and runoff into ground and surface waters.

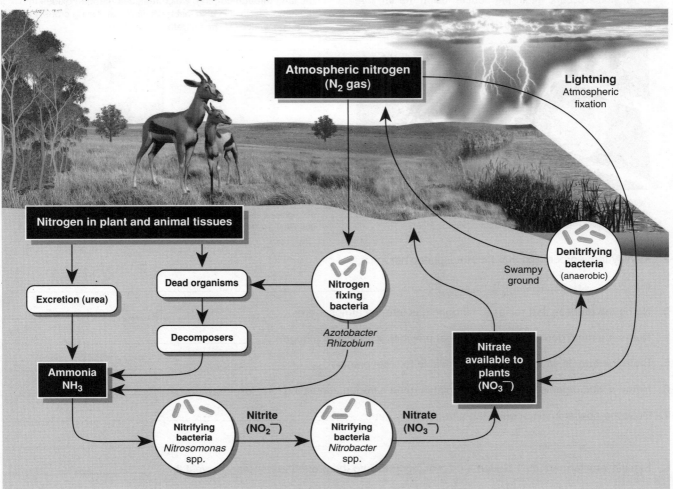

1. Describe five instances in the nitrogen cycle where **bacterial** action is important. Include the name of each of the processes and the changes to the form of nitrogen involved:

(a) _____

(b) _____

(c) _____

(d) _____

(e) _____

Image labels: Atmospheric nitrogen (N_2 gas); Lightning Atmospheric fixation; Nitrogen in plant and animal tissues; Excretion (urea); Dead organisms; Nitrogen fixing bacteria / Azotobacter Rhizobium; Denitrifying bacteria (anaerobic); Swampy ground; Decomposers; Ammonia NH_3; Nitrate available to plants (NO_3^-); Nitrifying bacteria Nitrosomonas spp.; Nitrite (NO_2^-); Nitrifying bacteria Nitrobacter spp.; Nitrate (NO_3^-).

Energy Flow and Nutrient Cycles

Code: RA 3

Nitrogen Fixation in Root Nodules

Root nodules are a root **symbiosis** between a higher plant and a bacterium. The bacteria fix atmospheric nitrogen and are extremely important to the nutrition of many plants, including the economically important legume family. Root nodules are extensions of the root tissue caused by entry of a bacterium. In legumes, this bacterium is *Rhizobium*. Other bacterial genera are involved in the root nodule symbioses in non-legume species.

The bacteria in these symbioses live in the nodule where they fix atmospheric nitrogen and provide the plant with most, or all, of its nitrogen requirements. In return, they have access to a rich supply of carbohydrate. The fixation of atmospheric nitrogen to ammonia occurs within the nodule, using the enzyme **nitrogenase**. Nitrogenase is inhibited by oxygen and the nodule provides a low O_2 environment in which fixation can occur.

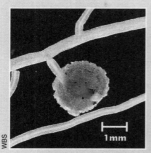 

Two examples of legume nodules caused by *Rhizobium*. The photographs above show the size of a single nodule (left), and the nodules forming clusters around the roots of *Acacia* (right).

Human Intervention in the Nitrogen Cycle

Until about sixty years ago, microbial nitrogen fixation (left) was the only mechanism by which nitrogen could be made available to plants. However, during WW II, Fritz Haber developed the **Haber process** whereby nitrogen and hydrogen gas are combined to form gaseous ammonia. The ammonia is converted into ammonium salts and sold as inorganic fertiliser. Its application has revolutionised agriculture by increasing crop yields.

As well as adding nitrogen fertilisers to the land, humans use anaerobic bacteria to break down livestock wastes and release NH_3 into the soil. They also intervene in the nitrogen cycle by discharging **effluent** into waterways. Nitrogen is removed from the land through burning, which releases nitrogen oxides into the atmosphere. It is also lost by mining, harvesting crops, and irrigation, which leaches nitrate ions from the soil.

Two examples of human intervention in the nitrogen cycle. The photographs above show the aerial application of a commercial fertiliser (left), and the harvesting of an agricultural crop (right).

2. Identify three processes that **fix** atmospheric nitrogen:

   (a) _____ (b) _____ (c) _____

3. Name the process that releases nitrogen gas into the atmosphere: _____

4. Name the main geological reservoir that provides a source of nitrogen: _____

5. State the form in which nitrogen is available to most plants: _____

6. Name a vital organic compound that plants need nitrogen containing ions for: _____

7. Describe how animals acquire the nitrogen they need: _____

8. Explain why farmers may plough a crop of legumes into the ground rather than harvest it: _____

9. Describe five ways in which humans may intervene in the nitrogen cycle and the effects of these interventions:

   (a) _____

   (b) _____

   (c) _____

   (d) _____

   (e) _____

The Phosphorus Cycle

Phosphorus is an essential component of nucleic acids and ATP. Unlike carbon, phosphorus has no atmospheric component; cycling of phosphorus is very slow and tends to be local. Small losses from terrestrial systems by leaching are generally balanced by gains from weathering. In aquatic and terrestrial ecosystems, phosphorus is cycled through food webs. Bacterial decomposition breaks down the remains of dead organisms and excreted products. Phosphatising bacteria further break down these products and return phosphates to the soil. Phosphorus is lost from ecosystems through run-off, precipitation, and sedimentation. Sedimentation may lock phosphorus away but, in the much longer term, it can become available again through processes such as geological uplift. Some phosphorus returns to the land as **guano**; phosphate-rich manure (typically of fish eating birds). This return is small though compared with the phosphate transferred to the oceans each year by natural processes and human activity. Excess phosphorus entering water bodies through runoff is a major contributor to **eutrophication** and excessive algal and weed growth, primarily because phosphorus is often limiting in aquatic systems.

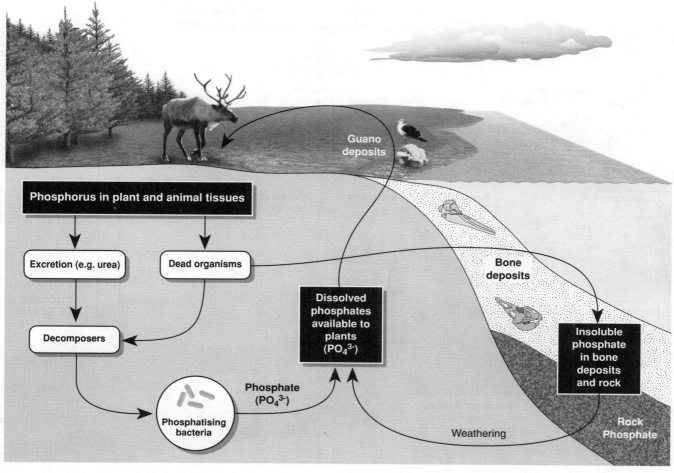

1. In the diagram, add an arrow and label to show where one human activity might intervene in the phosphorus cycle.

2. Identify two instances in the phosphorus cycle where bacterial action is important:

   (a) _____ (b) _____

3. Name two types of molecules found in living organisms which include phosphorus as a part of their structure:

   (a) _____ (b) _____

4. Name and describe the origin of three forms of inorganic phosphate making up the geological reservoir:

   (a) _____

   (b) _____

   (c) _____

5. Describe the processes that must occur in order to make rock phosphate available to plants again: _____
   _____

6. Identify one major difference between the phosphorus and carbon cycles: _____
   _____

Energy Flow and Nutrient Cycles

Code: ERA 2

The Water Cycle

The hydrologic cycle (water cycle), collects, purifies, and distributes the Earth's fixed supply of water. The main processes in this water recycling are described below. Besides replenishing inland water supplies, rainwater causes erosion and is a major medium for transporting dissolved nutrients within and among ecosystems. On a global scale, evaporation (conversion of water to gaseous water vapour) exceeds precipitation (rain, snow etc.) over the oceans. This results in a net movement of water vapour (carried by winds) over the land. On land, precipitation exceeds evaporation. Some of this precipitation becomes locked up in snow and ice, for varying lengths of time. Most forms surface and groundwater systems that flow back to the sea, completing the major part of the cycle. Living organisms, particularly plants, participate to varying degrees in the water cycle. Over the sea, most of the water vapour is due to evaporation alone. However on land, about 90% of the vapour results from plant transpiration. Animals (particularly humans) intervene in the cycle by utilising the resource for their own needs.

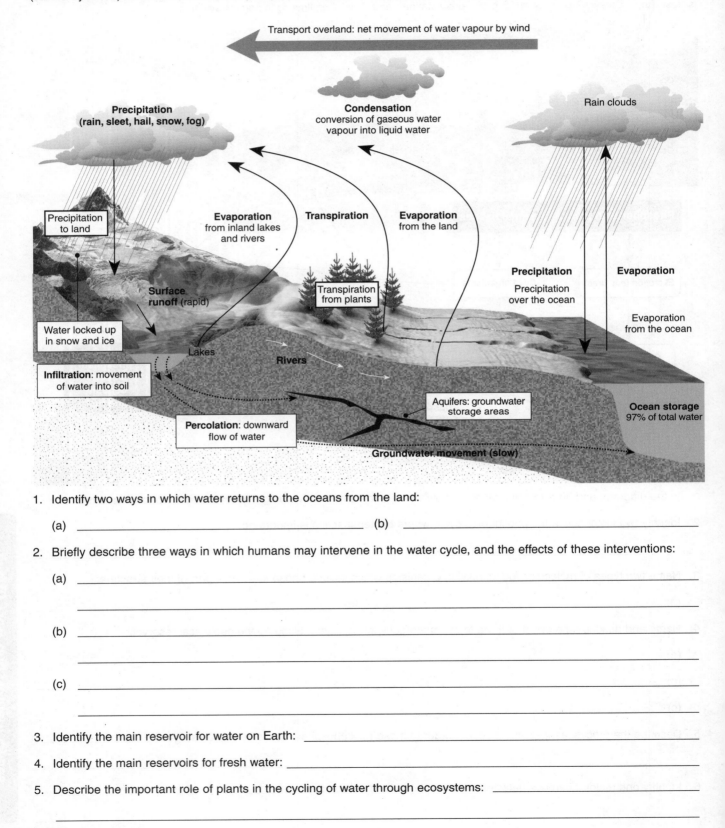

1. Identify two ways in which water returns to the oceans from the land:

   (a) _____ (b) _____

2. Briefly describe three ways in which humans may intervene in the water cycle, and the effects of these interventions:

   (a) _____

   _____

   (b) _____

   _____

   (c) _____

   _____

3. Identify the main reservoir for water on Earth: _____

4. Identify the main reservoirs for fresh water: _____

5. Describe the important role of plants in the cycling of water through ecosystems: _____

   _____

The Dynamics of Populations

Investigating the dynamics of populations

Features of populations, population growth and regulation, population age structure, r & K selection, intra- and interspecific interactions

Learning Objectives

☐ 1. Compile your own glossary from the **KEY WORDS** displayed in **bold type** in the learning objectives below.

Features of Populations *(pages 52-53, 61-62)*

☐ 2. Recall the difference between a **population** and a **community**. Explain what is meant by **population density** and distinguish it from **population size**.

☐ 3. Understand that populations are dynamic and exhibit attributes not shown by individuals themselves. Recognise the following attributes of populations: **population density**, population **distribution**, birth rate (**natality**), mean (average) age, death rate (**mortality**), **survivorship**, migration rate, average brood size, proportion of females breeding, **age structure**. Recognise that these attributes are population specific.

☐ 4. Describe, with examples, the distribution patterns of organisms within their range: uniform distribution, random distribution, clumped distribution. Suggest which factors govern each type of distribution.

Population Growth and Size *(pages 54-60)*

☐ 5. Recall that populations are dynamic. Outline how population size can be affected by **births**, **deaths**, and **migration** and express the relationship in an equation.

☐ 6. Recognise the value of **life tables** in providing information of patterns of population birth and mortality. Explain the role of **survivorship curves** in analysing populations. Providing examples, describe the features of Type I, II, and III survivorship curves. If required. distinguish between *r* and K selection.

☐ 7. Describe how the trends in population change can be shown in a **population growth curve** of population numbers (Y axis) against time (X axis).

☐ 8. Understand the factors that affect final population size, explaining clearly how they operate and providing examples where necessary. Include reference to:
   (a) **Carrying capacity** of the environment.
   (b) **Environmental resistance**.
   (c) **Density dependent factors**, e.g. intraspecific competition, interspecific competition, predation.
   (d) **Density independent factors**, e.g. climatic events.
   (e) **Limiting factors**, e.g. soil nutrient.

☐ 9. Distinguish between **exponential** and **sigmoidal growth curves**. Create labelled diagrams of these curves, indicating the different phases of growth and the factors regulating population growth at each stage.

☐ 10. Recognise patterns of population growth in colonising, stable, declining, and oscillating populations.

Species Interactions *(pages 63-72)*

☐ 11. Explain the nature of the **interspecific interactions** occurring in communities. Recognise: **competition**, **mutualism**, **commensalism**, **exploitation** (parasitism, predation, herbivory), **amensalism**, and **allelopathy**.

☐ 12. Describing at least one example, explain the possible effects of predator-prey interactions on the **population sizes** of both predator and prey.

☐ 13. Describe, and give examples of, **interspecific** and **intraspecific competition**. Explain the effects of **interspecific** and/or **intraspecific competition** on the distribution and/or population size of two species.

Supplementary Texts

See page 7 for additional details of these texts:
■ Allen, D, M. Jones, and G. Williams, 2001. **Applied Ecology**, pp.14-17.
■ Reiss, M. & J. Chapman, 2000. **Environmental Biology** (Cambridge University Press), pp. 3-5.
■ Smith, R. L. & T.M. Smith, 2001. **Ecology and Field Biology**, reading as required.

For web site references see the previous topic: *Energy Flow and Nutrient Cycles*

Periodicals

See page 7 for details of publishers of periodicals:

STUDENT'S REFERENCE

■ **The Other Side of Eden** Biol. Sci. Rev., 15(3) Feb. 2003, pp. 2-7. *An account of the Eden Project; its role in modelling ecosystem dynamics, including the interactions between species, is discussed.*

■ **Batesian Mimicry in Your Own Backyard** Biol. Sci. Rev., 17(3) Feb. 2005, pp. 25-27. *Batesian mimicry is seen in even the most common species.*

■ **Symbiosis: Mutual Benefit or Exploitation?** Biol. Sci. Rev., 7(4) March 1995, pp. 8-11. *An account of symbioses with illustrative examples.*

■ **Predator-Prey Relationships** Biol. Sci. Rev., 10(5) May 1998, pp. 31-35. *Predator-prey relationships, and the defence strategies of prey.*

■ **Inside Story** New Scientist, 29 April 2000, pp. 36-39. *Ecological interactions between fungi and plants and animals: what are the benefits?*

■ **The Future of Red Squirrels in Britain** Biol. Sci. Rev., 16(2) Nov. 2003, pp. 8-11. *A further account of the impact of the grey squirrel on Britain's native red squirrel populations.*

■ **Reds vs Grays: Squirrel Competition** Biol. Sci. Rev., 10(4) March 1998, pp. 30-31. *The nature of the competition between red and gray squirrels in the UK; an example of competitive exclusion?*

■ **Logarithms and Life** Biol. Sci. Rev., 13(4) March 2001, pp. 13-15. *The basics of logarithmic growth and its application to real populations.*

TEACHER'S REFERENCE

■ **Small Enclosures for Aquatic Ecology Experiments** The American Biology Teacher, 62 (6), June, 2000, pp. 424-428. *Using small aquatic populations to investigate life cycles, population dynamics, and community interactions.*

■ **Using Spreadsheets to Model Population Growth, Competition, and Predation in Nature** The American Biology Teacher, 61(4), April, 1999, pp. 294-296. *Using spreadsheets to improve understanding of models of population growth.*

Ecology

Presentation MEDIA to support this topic:

ECOLOGY
• Populations & Interactions

Features of Populations

Populations have a number of attributes that may be of interest. Usually, biologists wish to determine **population size** (the total number of organisms in the population). It is also useful to know the **population density** (the number of organisms per unit area). The density of a population is often a reflection of the **carrying capacity** of the environment, i.e. how many organisms an environment can support. Populations also have structure; particular ratios of different ages and sexes. These data enable us to determine whether the population is declining or increasing in size. We can also look at the **distribution** of organisms within their environment and so determine what particular aspects of the habitat are favoured over others. One way to retrieve information from populations is to **sample** them. Sampling involves collecting data about features of the population from samples of that population (since populations are usually too large to examine in total). Sampling can be carried out directly (by sampling the population itself using appropriate equipment) or indirectly (e.g. by monitoring calls or looking for droppings or other signs). Some of the population attributes that we can measure or calculate are illustrated on the diagram below.

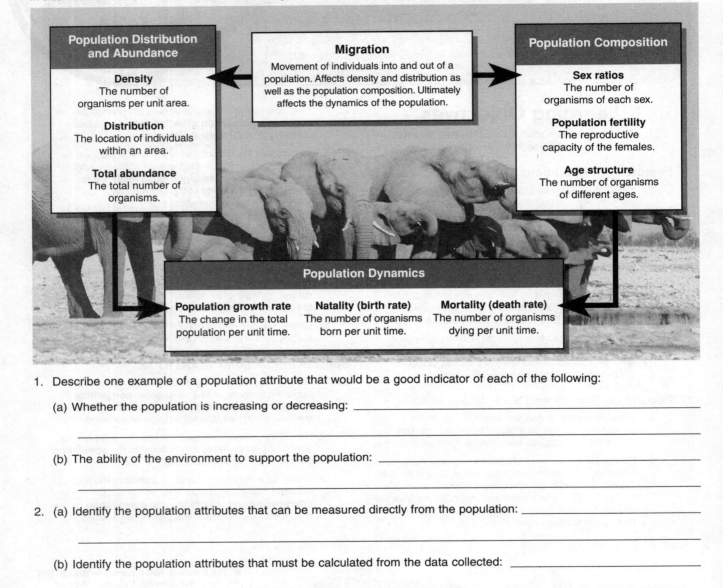

Population Distribution and Abundance

Density
The number of organisms per unit area.

Distribution
The location of individuals within an area.

Total abundance
The total number of organisms.

Migration
Movement of individuals into and out of a population. Affects density and distribution as well as the population composition. Ultimately affects the dynamics of the population.

Population Composition

Sex ratios
The number of organisms of each sex.

Population fertility
The reproductive capacity of the females.

Age structure
The number of organisms of different ages.

Population Dynamics

Population growth rate
The change in the total population per unit time.

Natality (birth rate)
The number of organisms born per unit time.

Mortality (death rate)
The number of organisms dying per unit time.

1. Describe one example of a population attribute that would be a good indicator of each of the following:

    (a) Whether the population is increasing or decreasing: _____

    _____

    (b) The ability of the environment to support the population: _____

    _____

2. (a) Identify the population attributes that can be measured directly from the population: _____

    _____

    (b) Identify the population attributes that must be calculated from the data collected: _____

    _____

3. Describe the value of population sampling for each of the following situations:

    (a) Conservation of a population of an endangered species: _____

    _____

    _____

    (b) Management of a fisheries resource: _____

    _____

    _____

Density and Distribution

Distribution and density are two interrelated properties of populations. Population density is the number of individuals per unit area (for land organisms) or volume (for aquatic organisms). Careful observation and precise mapping can determine the distribution patterns for a species. The three basic distribution patterns are: random, clumped and uniform. In the diagram below, the circles represent individuals of the same species. It can also represent populations of different species.

Low Density

In low density populations, individuals are spaced well apart. There are only a few individuals per unit area or volume (e.g. highly territorial, solitary mammal species).

High Density

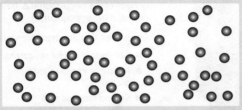

In high density populations, individuals are crowded together. There are many individuals per unit area or volume (e.g. colonial organisms, such as many corals).

Tigers are solitary animals, found at low densities.

Termites form well organised, high density colonies.

Random Distribution

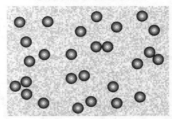 

Random distributions occur when the spacing between individuals is irregular. The presence of one individual does not directly affect the location of any other individual. Random distributions are uncommon in animals but are often seen in plants.

Clumped Distribution

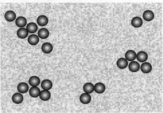 

Clumped distributions occur when individuals are grouped in patches (sometimes around a resource). The presence of one individual increases the probability of finding another close by. Such distributions occur in herding and highly social species.

Uniform Distribution

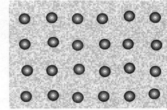 

Regular distribution patterns occur when individuals are evenly spaced within the area. The presence of one individual decreases the probability of finding another individual very close by. The penguins illustrated above are also at a high density.

1. Describe why some organisms may exhibit a clumped distribution pattern because of:

   (a) Resources in the environment: _____

   _____

   (b) A group social behaviour: _____

   _____

2. Describe a social behaviour found in some animals that may encourage a uniform distribution:

   _____

3. Describe the type of environment that would encourage uniform distribution:

   _____

4. Describe an example of each of the following types of distribution pattern:

   (a) Clumped: _____

   (b) Random (more or less): _____

   (c) Uniform (more or less): _____

Code: A 1

Population Regulation

Very few species show continued exponential growth. Population size is regulated by factors that limit population growth. The diagram below illustrates how population size can be regulated by environmental factors. **Density independent factors** may affect all individuals in a population equally. Some, however, may be better able to adjust to them. **Density dependent factors** have a greater affect when the population density is higher. They become less important when the population density is low.

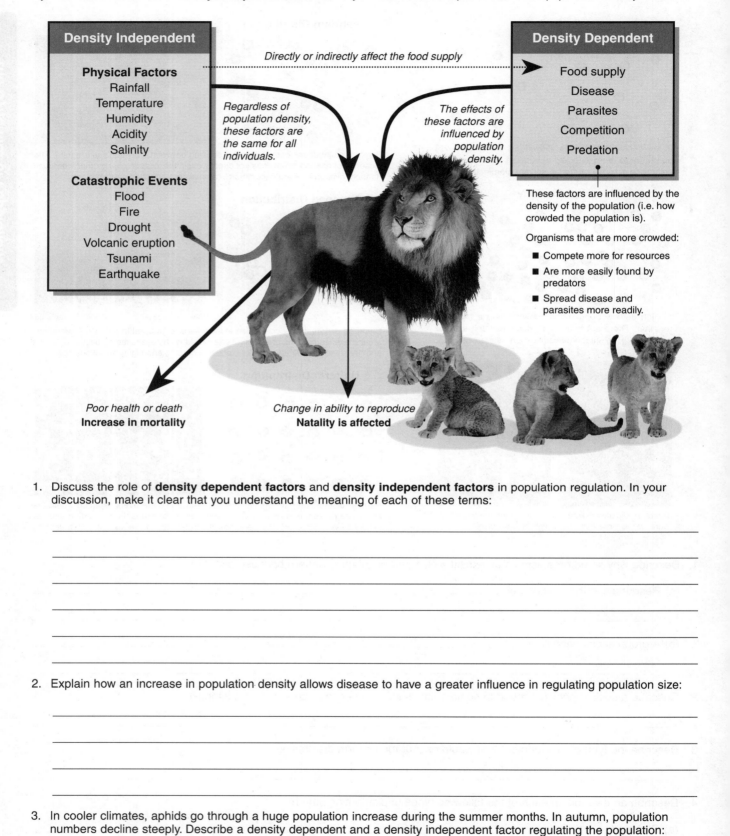

Density Independent

Physical Factors
Rainfall
Temperature
Humidity
Acidity
Salinity

Catastrophic Events
Flood
Fire
Drought
Volcanic eruption
Tsunami
Earthquake

Regardless of population density, these factors are the same for all individuals.

Directly or indirectly affect the food supply

The effects of these factors are influenced by population density.

Density Dependent
Food supply
Disease
Parasites
Competition
Predation

These factors are influenced by the density of the population (i.e. how crowded the population is).

Organisms that are more crowded:
■ Compete more for resources
■ Are more easily found by predators
■ Spread disease and parasites more readily.

Poor health or death
Increase in mortality

Change in ability to reproduce
Natality is affected

1. Discuss the role of **density dependent factors** and **density independent factors** in population regulation. In your discussion, make it clear that you understand the meaning of each of these terms:

   _____

   _____

   _____

   _____

   _____

   _____

2. Explain how an increase in population density allows disease to have a greater influence in regulating population size:

   _____

   _____

   _____

   _____

3. In cooler climates, aphids go through a huge population increase during the summer months. In autumn, population numbers decline steeply. Describe a density dependent and a density independent factor regulating the population:

   (a) Density dependent: _____

   (b) Density independent: _____

Population Growth

Organisms do not generally live alone. A **population** is a group of organisms of the same species living together in one geographical area. This area may be difficult to define as populations may comprise widely dispersed individuals that come together only infrequently (e.g. for mating). The number of individuals comprising a population may also fluctuate considerably over time. These changes make populations dynamic: populations gain individuals through births or immigration, and lose individuals through deaths and emigration. For a population in **equilibrium**, these factors balance out and there is no net change in the population abundance. When losses exceed gains, the population declines.

Births, deaths, immigrations (movements into the population) and *emigrations* (movements out of the population) are events that determine the numbers of individuals in a population. Population growth depends on the number of individuals added to the population from births and immigration, minus the number lost through deaths and emigration. This is expressed as:

> **Population growth =**
>
> **Births – Deaths + Immigration – Emigration**
> **(B)   (D)   (I)   (E)**

The difference between immigration and emigration gives *net migration*. Ecologists usually measure the **rate** of these events. These rates are influenced by environmental factors and by the characteristics of the organisms themselves. Rates in population studies are commonly expressed in one of two ways:

- Numbers per unit time, e.g. 20 150 live births per year.
- Per capita rate (number per head of population), e.g. 122 live births per 1000 individuals per year (12.2%).

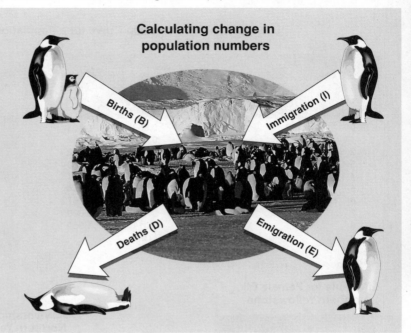

Calculating change in population numbers

Births (B) Immigration (I) Deaths (D) Emigration (E)

1. Define the following terms used to describe changes in population numbers:

   (a) Death rate (mortality): _____

   (b) Birth rate (natality): _____

   (c) Immigration: _____

   (d) Emigration: _____

   (e) Net migration rate: _____

2. Using the terms, B, D, I, and E (above), construct equations to express the following (the first is completed for you):

   (a) A population in equilibrium: _____ $B + I = D + E$ _____

   (b) A declining population: _____

   (c) An increasing population: _____

3. The rate of population change can be expressed as the interaction of all these factors:

   > Rate of population change = Birth rate – Death rate + Net migration rate (positive or negative)

   Using the formula above, determine the annual rate of population change for Mexico and the United States in 1972:

	USA	Mexico
Birth rate	1.73%	4.3%
Death rate	0.93%	1.0%
Net migration rate	+0.20%	0.0%

   Rate of population change for USA = _____

   Rate of population change for Mexico = _____

4. A population started with a total number of 100 individuals. Over the following year, population data were collected. Calculate birth rates, death rates, net migration rate, and rate of population change for the data below (as percentages):

   (a) Births = 14: Birth rate = _____ (b) Net migration = +2: Net migration rate = _____

   (c) Deaths = 20: Death rate = _____ (d) Rate of population change = _____

   (e) State whether the population is increasing or declining: _____

Life Tables & Survivorship

Life tables, such as those shown below, provide a summary of mortality for a population (usually for a group of individuals of the same age or **cohort**). The basic data are just the number of individuals remaining alive at successive sampling times (the **survivorship** or lx). Life tables are an important tool when analysing changes in populations over time. They can tell us the ages at which most mortality occurs in a population and can also provide information about life span and population age structure. From basic life table data, biologists derive survivorship curves, based on the lx column. Survivorship curves are standardised as the number of survivors per 1000 individuals so that populations of different types can be easily compared.

Life Table and Survivorship Curve for a Population of the Barnacle *Balanus*

Age in years (x)	No. alive each year (N_x)	Proportion surviving at the start of age x (l_x)	Proportion dying between x and x +1 (d_x)	Mortality (q_x)
0	142	1.000	0.563	0.563
1	62	0.437	0.198	0.452
2	34	0.239	0.098	0.412
3	20	0.141	0.035	0.250
4	15	0.106	0.028	0.267
5	11	0.078	0.036	0.454
6	6	0.042	0.028	0.667
7	2	0.014	0.0	0.000
8	2	0.014	0.014	1.000
9	0	0.0	0.0	–

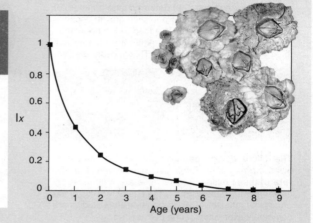

Life Table for Female Elk, Northern Yellowstone

x	l_x	d_x	q_x
0	1000	323	.323
1	677	13	.019
2	664	2	.003
3	662	2	.003
4	660	4	.006
5	656	4	.006
6	652	9	.014
7	643	3	.005
8	640	3	.005
9	637	9	.014
10	628	7	.001
11	621	12	.019
12	609	13	.021
13	596	41	.069
14	555	34	.061
15	521	20	.038
16	501	59	.118
17	442	75	.170
18	367	93	.253
19	274	82	.299
20	192	57	.297
21 +	135	135	1.000

Survivorship Curve for Female Elk of Northern Yellowstone National Park

1. (a) In the example of the barnacle *Balanus* above, state when most of the group die: _____

   (b) Identify the type of survivorship curve is represented by these data (see opposite): _____

2. (a) Using the grid, plot a survivorship curve for elk hinds (above) based on the life table data provided:

   (b) Describe the survivorship curve for these large mammals: _____
   _____

3. Explain how a biologist might use life table data to manage an endangered population: _____
   _____
   _____
   _____

Survivorship Curves

The survivorship curve depicts age-specific mortality. It is obtained by plotting the number of individuals of a particular cohort against time. Survivorship curves are standardised to start at 1000 and, as the population ages, the number of survivors progressively declines. The shape of a survivorship curve thus shows graphically at which life stages the highest mortality occurs. Survivorship curves in many populations fall into one of three hypothetical patterns (below). Wherever the curve becomes steep, there is an increase in mortality. The convex Type I curve is typical of populations whose individuals tend to live out their physiological life span. Such populations usually produce fewer young and show some degree of parental care. Organisms that suffer high losses of the early life stages (a Type III curve) compensate by producing vast numbers of offspring. These curves are conceptual models only, against which real life curves can be compared. Many species exhibit a mix of two of the three basic types. Some birds have a high chick mortality (Type III) but adult mortality is fairly constant (Type II). Some invertebrates (e.g. crabs) have high mortality only when moulting and show a stepped curve.

Hypothetical Survivorship Curves

Type I
Late loss survivorship curve
Mortality (death rate) is very low in the infant and juvenile years, and throughout most of adult life. Mortality increases rapidly in old age. **Examples**: Humans (in developed countries) and many other large mammals (e.g. big cats, elephants).

Type II
Constant loss survivorship curve
Mortality is relatively constant through all life stages (no one age is more susceptible than another). **Examples**: Some invertebrates such as *Hydra*, some birds, some annual plants, some lizards, and many rodents.

Type III
Early loss survivorship curve
Mortality is very high during early life stages, followed by a very low death rate for the few individuals reaching adulthood. **Examples**: Many fish (not mouth brooders) and most marine invertebrates (e.g. oysters, barnacles).

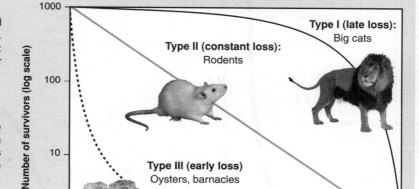

Graph of Survivorship in Relation to Age

Three basic types of survivorship curves and representative organisms for each type. The vertical axis may be scaled arithmetically or logarithmically.

Elephants have a close matriarchal society and a long period of parental care. Elephants are long-lived and females usually produce just one calf.

Rodents are well known for their large litters and prolific breeding capacity. Individuals are lost from the population at a more or less constant rate.

Despite vigilant parental care, many birds suffer high juvenile losses (Type III). For those surviving to adulthood, deaths occur at a constant rate.

1. Explain why human populations might not necessarily show a Type I curve: _____
_____

2. Explain how organisms with a Type III survivorship compensate for the high mortality during early life stages:
_____

3. Describe the features of a species with a Type I survivorship that aid in high juvenile survival: _____
_____

4. Discuss the following statement: "There is no standard survivorship curve for a given species; the curve depicts the nature of a population at a particular time and place and under certain environmental conditions.":
_____
_____
_____

Population Growth Curves

Populations becoming established in a new area for the first time are often termed **colonising populations** (below, left). They may undergo a rapid **exponential** (logarithmic) increase in numbers as there are plenty of resources to allow a high birth rate, while the death rate is often low. Exponential growth produces a J-shaped growth curve that rises steeply as more and more individuals contribute to the population increase. If the resources of the new habitat were endless (inexhaustible) then the population would continue to increase at an **exponential** rate. However, this rarely happens in natural populations. Initially, growth may be exponential (or nearly so), but as the population grows, its increase will slow and it will stabilise at a level that can be supported by the environment (called the carrying capacity or K). This type of growth is called sigmoidal and produces the **logistic growth curve** (below, right). **Established populations** will fluctuate about K, often in a regular way (grey area on the graph below, right). Some species will have populations that vary little from this stable condition, while others may oscillate wildly.

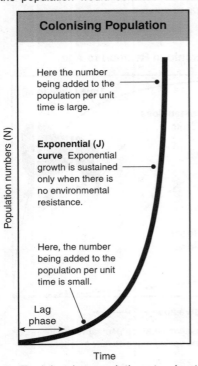

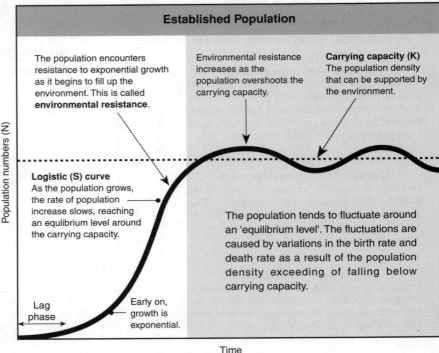

1. Explain why populations tend not to continue to increase exponentially in an environment: _____

_____

2. Explain what is meant by environmental resistance: _____

_____

3. (a) Explain what is meant by carrying capacity: _____

_____

   (b) Explain the importance of **carrying capacity** to the growth and maintenance of population numbers: _____

_____

4. Species that expand into a new area, such as rabbits did in areas of Australia, typically show a period of rapid population growth followed by a slowing of population growth as density dependent factors become more important and the population settles around a level that can be supported by the carrying capacity of the environment.

   (a) Explain why a newly introduced consumer (e.g. rabbit) would initially exhibit a period of exponential population growth:

_____

_____

   (b) Describe a likely outcome for a rabbit population after the initial rapid increase had slowed: _____

_____

5. Describe the effect that introduced grazing species might have on the carrying capacity of the environment:

_____

_____

Growth in a Bacterial Population

Bacteria normally reproduce by a process called **binary fission**; a simple mitotic cell division that is preceded by cell elongation and involves one cell dividing in two. The time required for a cell to divide is the **generation time** and it varies between organisms and with environmental conditions such as temperature. When a few bacteria are inoculated into a liquid growth medium, and the population is counted at intervals, it is possible to plot a

bacterial **growth curve** that shows the growth of cells over time. In this activity, you will simulate this for a hypothetical bacterial population with a generation time of 20 minutes. In a bacterial culture with a limited nutrient supply, four growth phases are evident: the early **lag phase**, the **log phase** of exponential growth, the **stationary phase** when growth rate slows, and the **death** phase, when the population goes into logarithmic decline.

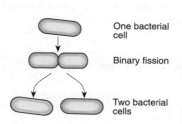

One bacterial cell

Binary fission

Two bacterial cells

Time (mins)	Population size
0	1
20	2
40	4
60	8
80	
100	
120	
140	
160	
180	
200	
220	
240	
260	
280	
300	
320	
340	
360	

1. Complete the table (above) by doubling the number of bacteria for every 20 minute interval.

2. Graph the results on the graph grid above. Make sure that you choose suitable scales for each axis. Label the axes and mark out (number) the scale for each axis. Identify the **lag** and **log** phases of growth and mark them on the graph.

3. State how many bacteria were present after: 1 hour: _____   3 hours: _____   6 hours: _____

4. Describe the shape of the curve you have plotted: _____

5. Predict what would happen to the shape of the growth curve of this population assuming no further input of nutrients:

_____

_____

Code: DA 2

r and K Selection

Two parameters govern the logistic growth of populations: the intrinsic rate of natural increase or biotic potential (this is the maximum reproductive potential of an organism, symbolised by an italicised r), and the carrying capacity (saturation density) of the environment (represented by the letter **K**). Species can be characterised by the relative importance of r and K in their life cycles. Species with a high intrinsic capacity for population increase are called **r-selected species**, and include algae, bacteria, rodents, many insects, and most annual plants. These species show life history features associated with rapid growth in disturbed environments. To survive, they must continually invade new areas to compensate for being replaced by more competitive species. In contrast, **K-selected** species, which include most large mammals, birds of prey, and large, long-lived plants, exist near the carrying capacity of their environments and are pushed in competitive environments to use resources more efficiently. These species have fewer offspring and longer lives, and put their energy into nuturing their young to reproductive age. Most organisms have reproductive patterns between these two extremes. Both r-selected species (crops) and K-selected species (livestock) are found in agriculture.

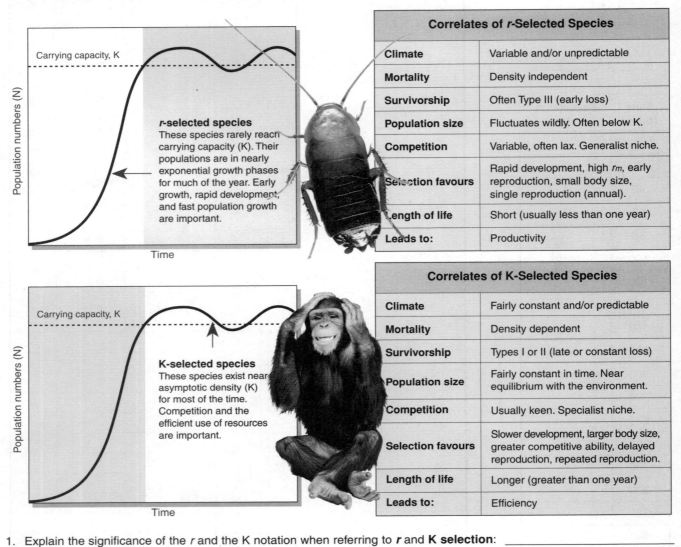

Correlates of r-Selected Species

Climate	Variable and/or unpredictable
Mortality	Density independent
Survivorship	Often Type III (early loss)
Population size	Fluctuates wildly. Often below K.
Competition	Variable, often lax. Generalist niche.
Selection favours	Rapid development, high r_m, early reproduction, small body size, single reproduction (annual).
Length of life	Short (usually less than one year)
Leads to:	Productivity

r-selected species
These species rarely reach carrying capacity (K). Their populations are in nearly exponential growth phases for much of the year. Early growth, rapid development, and fast population growth are important.

Correlates of K-Selected Species

Climate	Fairly constant and/or predictable
Mortality	Density dependent
Survivorship	Types I or II (late or constant loss)
Population size	Fairly constant in time. Near equilibrium with the environment.
Competition	Usually keen. Specialist niche.
Selection favours	Slower development, larger body size, greater competitive ability, delayed reproduction, repeated reproduction.
Length of life	Longer (greater than one year)
Leads to:	Efficiency

K-selected species
These species exist near asymptotic density (K) for most of the time. Competition and the efficient use of resources are important.

1. Explain the significance of the r and the K notation when referring to r and **K selection**: _____

_____

2. Giving an example, explain why r-selected species tend to be **opportunists**: _____

_____

_____

3. Explain why K-selected species are also called **competitor species**: _____

_____

_____

4. Suggest why many K-selected species are often vulnerable to extinction: _____

_____

Population Age Structure

The **age structure** of a population refers to the relative proportion of individuals in each age group in the population. The age structure of populations can be categorised according to specific age categories (such as years or months), but also by other measures such as life stage (egg, larvae, pupae, instars), of size class (height or diameter in plants). Population growth is strongly influenced by age structure; a population with a high proportion of reproductive and prereproductive aged individuals has a much greater potential for population growth than one that is dominated by older individuals. The ratio of young to adults in a relatively stable population of most mammals and birds is approximately 2:1 (below, left). Growing populations in general are characterised by a large and increasing number of young, whereas a population in decline typically has a decreasing number of young. Population age structures are commonly represented as pyramids, in which the proportions of individuals in each age/size class are plotted with the youngest individuals at the pyramid's base. The number of individuals moving from one age class to the next influences the age structure of the population from year to year. The loss of an age class (e.g. through overharvesting) can profoundly influence a population's viability and can even lead to population collapse.

The Dynamics of Populations

Age Structures in Animal Populations

These theoretical age pyramids, which are especially applicable to birds and mammals, show how growing populations are characterised by a high ratio of young (white bar) to adult age classes (grey bars). Ageing populations with poor production are typically dominated by older individuals.

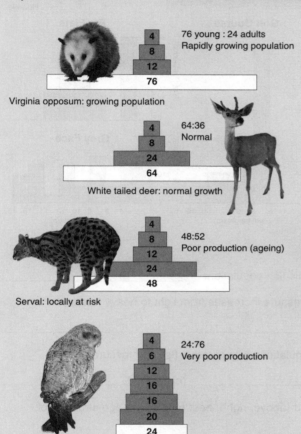

Virginia opposum: growing population

4	76 young : 24 adults
8	Rapidly growing population
12	
76	

White tailed deer: normal growth

4	64:36
8	Normal
24	
64	

Serval: locally at risk

4	48:52
8	Poor production (ageing)
12	
24	
48	

Kakapo: endangered

4	24:76
6	Very poor production
12	
16	
16	
20	
24	

Age Structures in Human Populations

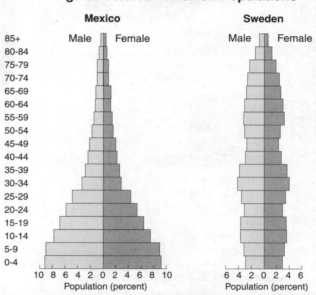

Mexico — Male / Female

Sweden — Male / Female

Age classes: 85+, 80-84, 75-79, 70-74, 65-69, 60-64, 55-59, 50-54, 45-49, 40-44, 35-39, 30-34, 25-29, 20-24, 15-19, 10-14, 5-9, 0-4

Mexico Population (percent): 10 8 6 4 2 0 2 4 6 8 10

Sweden Population (percent): 6 4 2 0 2 4 6

Extended family: Samoa

Most of the growth in human populations in recent years has occurred in the developing countries in Africa, Asia, and Central and South America. This is reflected in their age structure; a large proportion of the population comprises individuals younger than 15 years (age pyramid above, left). Even if each has fewer children, the population will continue to increase for many years. The stable age structure of Sweden is shown for comparison.

1. For the theoretical age pyramids above left:

   (a) State the approximate ratio of young to adults in a rapidly increasing population: _____

   (b) Suggest why changes in population age structure alone are not necessarily a reliable predictor of population trends:

   _____

   _____

2. Explain why the population of Mexico is likely to continue to increase rapidly even if the rate of population growth slows:

   _____

   _____

Code: RDA 2

Analysis of the age structure of a population can assist in its management because it can indicate where most of the mortality occurs and whether or not reproductive individuals are being replaced. The age structure of both plant and animal populations can be examined; a common method is through an analysis of size which is often related to age in a predictable way.

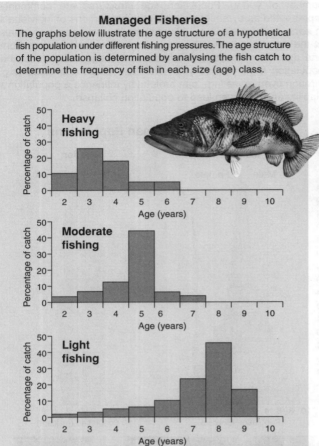

Managed Fisheries

The graphs below illustrate the age structure of a hypothetical fish population under different fishing pressures. The age structure of the population is determined by analysing the fish catch to determine the frequency of fish in each size (age) class.

Thatch Palm Populations on Lord Howe Island

Lord Howe Island is a narrow sliver of land approximately 770 km northeast of Sydney. The age structure of populations of the thatch palm *Howea forsteriana* was determined at three locations on the island: the golf course, Grey Face and Far Flats. The height of the stem was used as an indication of age. The differences in age structure between the three sites are mainly due to the extent of grazing at each site.

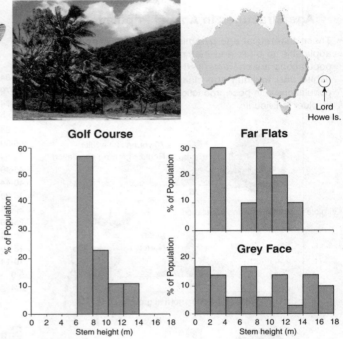

3. For the managed fish population above left:

(a) Name the general factor that changes the age structure of this fish population: _____

(b) Describe how the age structure changes when the fishing pressure increases from light to heavy levels:

_____

4. State the most common age class for each of the above fish populations with different fishing pressures:

(a) Heavy: _____ (b) Moderate: _____ (c) Light: _____

5. Determine which of the three sites sampled on Lord Howe Island (above, right), best reflects the age structure of:

(a) An ungrazed population: _____

Reason for your answer: _____

(b) A heavily grazed and mown population: _____

Reason for your answer: _____

6. Describe the likely long term prospects for the population at the golf course: _____

_____

7. Describe a potential problem with using size to estimate age: _____

_____

8. Explain why a knowledge of age structure could be important in managing a resource: _____

_____

Species Interactions

No organism exists in isolation. Each takes part in many interactions, both with other organisms and with the non-living components of the environment. Species interactions may involve only occasional or indirect contact (predation or competition) or they may involve close association or **symbiosis**. Symbiosis is a term that encompasses a variety of interactions involving close species contact. There are three types of symbiosis: **parasitism** (a form of exploitation), **mutualism**, and **commensalism**. Species interactions affect population densities and are important in determining community structure and composition. Some interactions, such as allelopathy, may even determine species presence or absence in an area.

The Dynamics of Populations

Examples of Species Interactions

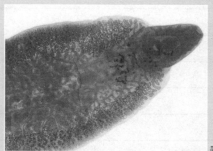

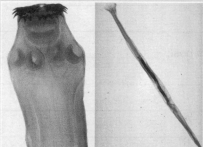

Parasitism is a common exploitative relationship in plants and animals. A parasite exploits the resources of its host (e.g. for food, shelter, warmth) to its own benefit. The host is harmed, but usually not killed. **Endoparasites**, such as liver flukes (left), tapeworms (centre) and nematodes (right)), are highly specialised to live inside their hosts, attached by hooks or suckers to the host's tissues.

Ectoparasites, such as ticks (above), mites, and fleas, live attached to the outside of the host, where they suck body fluids, cause irritation, and may act as vectors for disease causing microorganisms.

Mutualism involves an intimate association between two species that offers advantage to both. **Lichens** (above) are the result of a mutualism between a fungus and an alga (or cyanobacterium).

Termites have a mutualistic relationship with the cellulose digesting bacteria in their guts. A similar mutualistic relationship exists between ruminants and their gut microflora of bacteria and ciliates.

In **commensal** relationships, such as between this large grouper and a remora, two species form an association where one organism, the commensal, benefits and the other is neither harmed or helped.

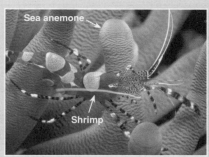

Many species of decapod crustaceans, such as this anemone shrimp, are commensal with sea anemones. The shrimp gains by being protected from predators by the anemone's tentacles.

Interactions involving **competition** for the same food resources are dominated by the largest, most aggressive species. Here, hyaenas compete for a carcass with vultures and maribou storks.

Predation is an easily identified relationship, as one species kills and eats another (above). Herbivory is similar type of exploitation, except that the plant is usually not killed by the herbivore.

1. Discuss each of the following interspecific relationships, including reference to the species involved, their role in the interaction, and the specific characteristics of the relationship:

   (a) **Mutualism** between ruminant herbivores and their gut microflora: _____

   _____

   _____

   _____

   _____

(b) **Commensalism** between a shark and a remora: _____

_____

_____

_____

(c) **Parasitism** between a tapeworm and its human host: _____

_____

_____

_____

(d) **Parasitism** between a cat flea and its host: _____

_____

_____

_____

2. Summarise your knowledge of species interactions by completing the following, entering a (+), (–), or (0) for species B, and writing a brief description of each term. Codes: (+): species benefits, (–): species is harmed, (0): species is unaffected.

Interaction	Species A	Species B	Description of relationship
(a) Mutualism	+		
(b) Commensalism	+		
(c) Parasitism	–		
(d) Amensalism	0		
(e) Predation	–		
(f) Competition	–		
(g) Herbivory	+		
(h) Antibiosis	+ / 0		

3. For each of the interactions between two species described below, choose the correct term to describe the interaction and assign a +, – or 0 for each species involved in the space supplied. Use the completed table above to help you:

Description	Term	Species A	Species B
(a) A tiny cleaner fish picking decaying food from the teeth of a much larger fish (e.g. grouper).	Mutualism	Cleaner fish +	Grouper +
(b) Ringworm fungus growing on the skin of a young child.		Ringworm	Child
(c) Human effluent containing poisonous substances killing fish in a river downstream of discharge.		Humans	Fish
(d) Humans planting cabbages to eat only to find that the cabbages are being eaten by slugs.		Humans	Slugs
(e) A shrimp that gets food scraps and protection from sea anemones, which appear to be unaffected.		Shrimp	Anemone
(f) Birds follow a herd of antelopes to feed off disturbed insects, antelopes alerted to danger by the birds.		Birds	Antelope

Predator-Prey Strategies

A predator eating its prey is one of the most conspicuous species interactions. In most cases, the predator and prey are different species, though cannibalism is quite common in some species. Predators have acute senses with which to locate and identify prey. Many also have structures such as teeth, claws, and poison to catch and subdue their prey. Animals can avoid being eaten by using passive defences, such as hiding, or active ones, such as rapid escape or aggressive defence.

Predator Avoidance Strategies Among Animals

Batesian mimicry

Common wasp

Wasp beetle

Harmless prey gain immunity from attack by mimicking harmful animals (called **Batesian mimicry**). The beetle on the right has the same black and yellow colour scheme as the common wasps, which can give a painful sting.

Müllerian mimicry

Monarch butterfly

Queen butterfly

Many unpalatable species tend to resemble each other; a phenomenon known as **Müllerian mimicry**. The monarch and queen butterflies are both toxic. By looking alike, these mimics present a common image for predators to avoid.

Poisonous

Arrow poison frog

Lionfish

Highly toxic animals, such as arrow poison frogs and lionfish, may advertise this fact with bright colours. The evolution of toxicity may involve kin selection, as the parent may have to die to educate a future predator of its young.

Chemical defence

Skunk

Some animals can produce offensive smelling chemicals. American skunks squirt a nauseous fluid at attackers.

Hiding

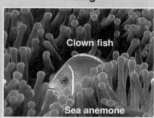
Clown fish

Sea anemone

Animals, such as this clown fish, take refuge in the safety provided by animals with more effective defence.

Detection and escape

Grasshopper

Some animals, like this grasshopper, are masters of detecting approaching danger and making a rapid escape.

Camouflage

Leaf insect

Cryptic shape and coloration allows some animals, such as this leaf insect, to blend into their background.

Startle display

Stick insect

Frill necked lizard

Effective startle displays must come as a surprise. This stick insect will rear up on its hind legs and fan its wings. The frill necked lizard, native to Australia, will rear up suddenly and erect its frill before retreating off at high speed.

Visual deception

Owl butterfly

Butterfly fish

Deceptive markings such as large, fake eyes can apparently deceive predators momentarily, allowing the prey to escape. They may also induce the predator to strike at a nonvital end of the prey. Note the fish's real eye is disguised.

Group defence

Fish often swim in large schools, which move together as one mass in a way that confuses predators.

When birds, such as these flamingoes, form large flocks, the risk of predation to each individual is reduced.

Armoured defence

Tortoise

Pill millipede

Tough outer cases that hamper predators are common in both vertebrate and invertebrate prey. Almost all molluscs have protective shells built around their sensitive parts. Some animals can coil up to protect vulnerable parts.

Offensive weapons

Burr fish

Deer

Offensive weapons are essential if prey are to actively fend off an attack by a predator. Many grazing mammals have sharp horns and can repel attacks by predators. Some animal bodies have spikes that offer passive defence.

Code: RA 3

Prey Capturing Strategies

Concealment	Filter feeding	Lures	Traps

Praying mantis

Manta ray

Angler fish

Web spider

Some animals camouflage themselves in their surroundings, striking when the prey comes within reach.

Many marine animals (e.g. barnacles, baleen whales, sponges, manta rays) filter the water to extract tiny plankton.

This angler fish, glow worms and a type of spider all use lures to attract prey within striking range.

Spiders have developed a unique method of trapping their prey. Strong, sticky silk threads trap flying insects.

Tool use	Stealth	Speed	Group attack

Chimpanzee

Rattlesnake

Infrared pit

Cheetah

Pelicans

Some animals are gifted tool users. Chimpanzees use carefully prepared twigs to extract termites from mounds.

The night hunting ability of some poisonous snakes is greatly helped by the presence of infrared senses.

Some animals, such as cheetahs and some predatory birds, can simply outrun or outfly their prey.

Cooperative group behaviour may make prey capture much easier. Pelicans herd fish into 'killing zones'.

1. Explain why poisonous (unpalatable) animals are often brightly coloured so that they are easily seen:

2. Describe the purpose of large, fake eyes on some butterflies and fish:

3. Describe a behaviour typical of a (named) prey species that makes them difficult to detect by a predator:

4. Describe a behaviour of prey that is actively defensive:

5. Describe two behaviours of (named) predators that facilitate prey capture:

   (a)

   (b)

6. Discuss **Batesian and Müllerian mimicry**, explaining the differences between them and any benefits to the species involved directly or indirectly in the relationship:

7. Describe two possible ways in which toxicity could evolve in prey species, given the prey has to be eaten in order for the predator to learn from the experience:

   (a)

   (b)

Niche Differentiation

Competition is most intense between members of the same species because their habitat and resource requirements are identical. Interspecific competition (between different species) is often less intense. Species with similar ecological requirements may reduce competition by exploiting microhabitats within the ecosystem. In the eucalypt forest below, different bird species exploit tree trunks, leaf litter, different levels within the canopy, and air space. Competition may also be reduced by exploiting the same resources at a different time of the day or year.

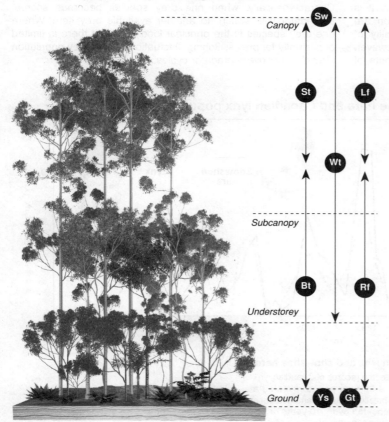

Reducing competition in a eucalypt forest

The diagram on the left shows the foraging heights of birds in an eastern Australian eucalypt forest. A wide variety of food resources are offered by the structure of the forest. Different layers of the forest allow birds to specialise in foraging at different heights. The ground-dwelling yellow-throated scrubwren and ground thrush have robust legs and feet, while the white-throated treecreeper has long toes and large curved claws and the swifts are extremely agile fliers capable of catching insects on the wing.

Key to bird species

Ys	Yellow-throated scrubwren	**Lf**	Leaden flycatcher
Bt	Brown thornbill	**Gt**	Ground thrush
Sw	Spine-tailed swift	**Rf**	Rufous fantail
St	Striated thornbill	**Wt**	White-throated treecreeper

Adapted from: Recher *et al.*, 1986. *A Natural Legacy: Ecology in Australia.* Maxwell Macmillan Publishing Australia.

Distribution of ecologically similar fish

The diagram below shows the distribution of ecologically similar damsel fish over a coral reef at Heron Island, Queensland, Australia. The habitat and resource requirements of these species overlap considerably.

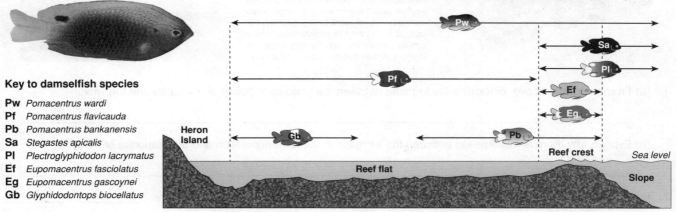

Key to damselfish species

Pw *Pomacentrus wardi*
Pf *Pomacentrus flavicauda*
Pb *Pomacentrus bankanensis*
Sa *Stegastes apicalis*
Pl *Plectroglyphidodon lacrymatus*
Ef *Eupomacentrus fasciolatus*
Eg *Eupomacentrus gascoynei*
Gb *Glyphidodontops biocellatus*

1. Describe two ways in which species can avoid directly competing for the same resources in their habitat:

   (a) _____

   (b) _____

2. Explain why intraspecific competition is more intense than interspecific competition: _____

   _____

3. Suggest how the damsel fish on the reef at Heron Island (above) might reduce competition: _____

   _____

   _____

The Dynamics of Populations

Code: A 2

Predator-Prey Interactions

Some mammals, particularly in highly seasonal environments, exhibit regular cycles in their population numbers. Snowshoe hares in Canada exhibit such a cycle of population fluctuation that has a periodicity of 9–11 years. Populations of lynx in the area show a similar periodicity. Contrary to early suggestions that the lynx controlled the size of the hare population, it is now known that the fluctuations in the hare population are governed by other factors, most probably the availability of palatable grasses. The fluctuations in the lynx numbers however, do appear to be the result of fluctuations in the numbers of

hares (their principal food item). This is true of most **vertebrate** predator-prey systems: predators do not usually control prey populations, which tend to be regulated by other factors such as food availability and climatic factors. Most predators have more than one prey species, although one species may be preferred. Characteristically, when one prey species becomes scarce, a predator will "switch" to another available prey item. Where one prey species is the principal food item and there is limited opportunity for prey switching, fluctuations in the prey population may closely govern predator cycles.

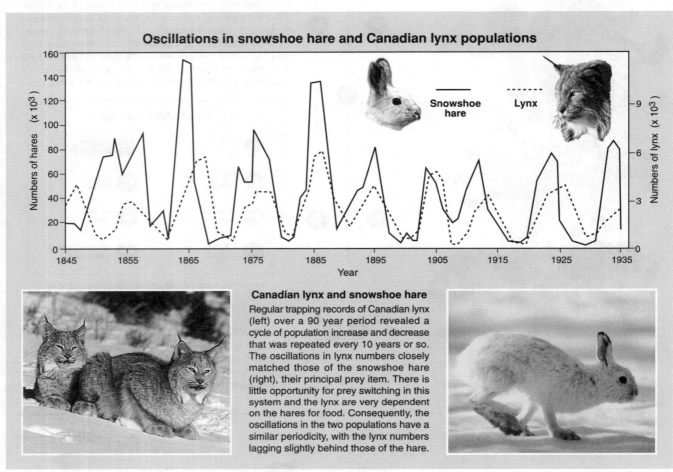

Oscillations in snowshoe hare and Canadian lynx populations

Canadian lynx and snowshoe hare
Regular trapping records of Canadian lynx (left) over a 90 year period revealed a cycle of population increase and decrease that was repeated every 10 years or so. The oscillations in lynx numbers closely matched those of the snowshoe hare (right), their principal prey item. There is little opportunity for prey switching in this system and the lynx are very dependent on the hares for food. Consequently, the oscillations in the two populations have a similar periodicity, with the lynx numbers lagging slightly behind those of the hare.

1. (a) From the graph above, determine the lag time between the population peaks of the hares and the lynx:

_____

(b) Explain why there is this time lag between the increase in the hare population and the response of the lynx:

_____

2. Suggest why the lynx populations appear to be so dependent on the fluctuations on the hare: _____

_____

_____

3. (a) In terms of birth and death rates, explain how the availability of palatable food might regulate the numbers of hares:

_____

_____

(b) Explain how a decline in available palatable food might affect their ability to withstand predation pressure:

_____

_____

Interspecific Competition

In naturally occurring populations, direct competition between different species (**interspecific competition**) is usually less intense than intraspecific competition because coexisting species have evolved slight differences in their realised niches, even though their fundamental niches may overlap (a phenomenon termed **niche differentiation**). However, when two species with very similar niche requirements are brought into direct competition through the introduction of a foreign species, one usually benefits at the expense of the other. The inability of two species with the same described niche to coexist is referred to as the **competitive exclusion principle**. In Britain, introduction of the larger, more aggressive, grey squirrel in 1876 has contributed to a contraction in range of the native red squirrel (below), and on the Scottish coast, this phenomenon has been well documented in barnacle species (opposite). The introduction of ecologically aggressive species is often implicated in the displacement or decline of native species, although there may be more than one contributing factor. Displacement of native species by introduced ones is more likely if the introduced competitor is also adaptable and hardy. It can be difficult to provide evidence of decline in a species as a direct result of competition, but it is often inferred if the range of the native species contracts and that of the introduced competitor shows a corresponding increase.

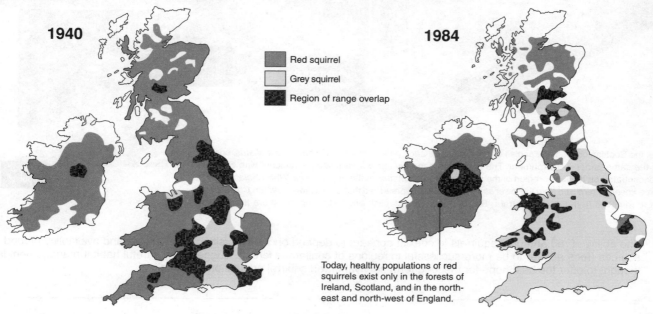

1940

1984

Red squirrel
Grey squirrel
Region of range overlap

Today, healthy populations of red squirrels exist only in the forests of Ireland, Scotland, and in the north-east and north-west of England.

Red squirrel

The **European red squirrel**, *Sciurus vulgaris*, was the only squirrel species in Britain until the introduction of the **American grey squirrel**, *Sciurus carolinesis*, in 1876. In 44 years since the 1940 distribution survey (above left), the more adaptable grey squirrel has displaced populations of the native red squirrels over much of the British Isles, particularly in the south (above right). Whereas the red squirrels once occupied both coniferous and broad leafed woodland, they are now almost solely restricted to coniferous forest and are completely absent from much of their former range.

Grey squirrel

1. Outline the evidence to support the view that the red-grey squirrel distributions in Britain are an example of the competitive exclusion principle:

_____

_____

_____

2. Some biologists believe that competition with grey squirrels is only one of the factors contributing to the decline in the red squirrels in Britain. Explain the evidence from the 1984 distribution map that might support this view:

_____

_____

_____

_____

Competitive Exclusion in Barnacles

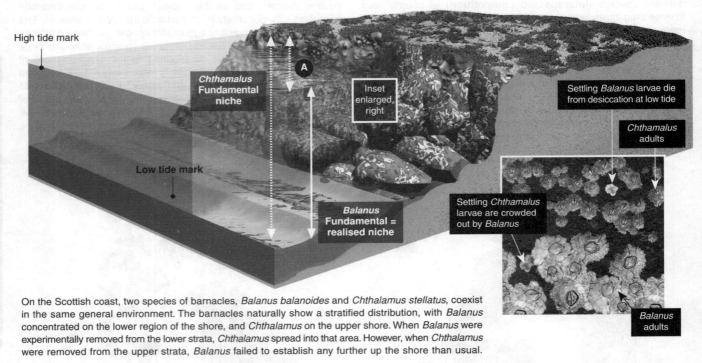

High tide mark

Chthamalus
Fundamental
niche

A

Inset
enlarged,
right

Low tide mark

Balanus
Fundamental =
realised niche

Settling *Balanus* larvae die
from desiccation at low tide

Chthamalus
adults

Settling *Chthamalus*
larvae are crowded
out by *Balanus*

Balanus
adults

On the Scottish coast, two species of barnacles, *Balanus balanoides* and *Chthalamus stellatus*, coexist in the same general environment. The barnacles naturally show a stratified distribution, with *Balanus* concentrated on the lower region of the shore, and *Chthalamus* on the upper shore. When *Balanus* were experimentally removed from the lower strata, *Chthalamus* spread into that area. However, when *Chthalamus* were removed from the upper strata, *Balanus* failed to establish any further up the shore than usual.

3. The ability of red and grey squirrels to coexist appears to depend on the diversity of habitat type and availability of food sources (reds appear to be more successful in regions of coniferous forest). Suggest why careful habitat management is thought to offer the best hope for the long term survival of red squirrel populations in Britain:

_____
_____
_____
_____
_____
_____

4. Suggest other conservation methods that might aid the survival of viable red squirrel populations:

_____
_____
_____

5. (a) In the example of the barnacles (above), describe what is represented by the zone labelled with the arrow **A**:

_____

(b) Outline the evidence for the barnacle distribution being the result of competitive exclusion: _____

_____
_____

6. Describe two aspects of the biology of a named introduced species that have helped its success as an invading competitor:

Species: _____

(a) _____

(b) _____

Intraspecific Competition

Some of the most intense competition occurs between individuals of the same species (**intraspecific competition**). Most populations have the capacity to grow rapidly, but their numbers cannot increase indefinitely because environmental resources are finite. Every ecosystem has a **carrying capacity** (K), defined as the number of individuals in a population that the environment can support. Intraspecific competition for resources increases with increasing population size and, at carrying capacity, it reduces the per capita growth rate to zero. When the demand for a particular resource (e.g. food, water, nesting sites, nutrients, or light) exceeds supply, that resource becomes a **limiting factor**. Populations respond to resource limitation by reducing their population growth rate (e.g. through lower birth rates or higher mortality). The response of individuals to limited resources varies depending on the organism. In many invertebrates and some vertebrates such as frogs, individuals reduce their growth rate and mature at a smaller size. In some vertebrates, territoriality spaces individuals apart so that only those with adequate resources can breed. When resources are very limited, the number of available territories will decline.

Intraspecific Competition

Scramble competition in caterpillars

Contest competition in wolves

Display of a male anole

Direct competition for available food between members of the same species is called **scramble competition**. In some situations where scramble competition is intense, none of the competitors gets enough food to survive.

In some cases, competition is limited by hierarchies existing within a social group. Dominant individuals receive adequate food, but individuals low in the hierarchy must **contest** the remaining resources and may miss out.

Intraspecific competition may be for mates or breeding sites, as well as for food. In anole lizards (above), males have a bright red throat pouch and use much of their energy displaying to compete with other males for available mates.

Competition Between Tadpoles of *Rana tigrina*

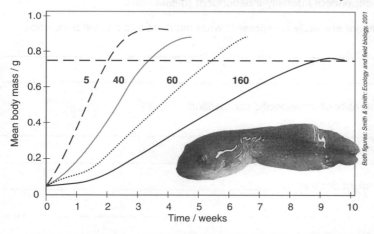

Both figures: Smith & Smith: Ecology and field biology, 2001

Food shortage reduces both individual growth rate and survival, and population growth. In some organisms, where there is a metamorphosis or a series of moults before adulthood (e.g. frogs, crustacean zooplankton, and butterflies), individuals may die before they mature.

The graph (left) shows how the growth rate of tadpoles (*Rana tigrina*) declines as the density increases from 5 to 160 individuals (in the same sized space).

- At high densities, tadpoles grow more slowly, taking longer to reach the minimum size for metamorphosis (0.75 g), and decreasing their chances of successfully metamorphosing from tadpoles into frogs.
- Tadpoles held at lower densities grow faster, to a larger size, metamorphosing at an average size of 0.889 g.
- In some species, such as frogs and butterflies, the adults and juveniles reduce the intensity of intraspecific competition by exploiting different food resources.

1. Using an example, predict the likely effects of **intraspecific competition** on each of the following:

(a) Individual growth rate: _____

_____

_____

(b) Population growth rate: _____

_____

_____

(c) Final population size: _____

_____

_____

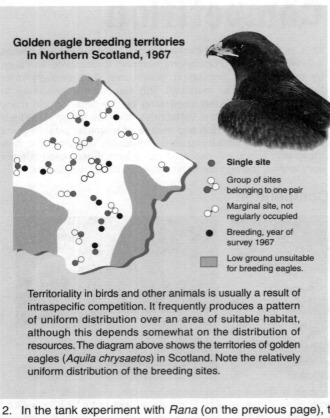

Golden eagle breeding territories in Northern Scotland, 1967

- ● Single site
- ◐ Group of sites belonging to one pair
- ○ Marginal site, not regularly occupied
- ● Breeding, year of survey 1967
- ▨ Low ground unsuitable for breeding eagles.

Territoriality in birds and other animals is usually a result of intraspecific competition. It frequently produces a pattern of uniform distribution over an area of suitable habitat, although this depends somewhat on the distribution of resources. The diagram above shows the territories of golden eagles (*Aquila chrysaetos*) in Scotland. Note the relatively uniform distribution of the breeding sites.

Territoriality in Great Tits (*Parus major*)

Six breeding pairs of great tits were removed from an oak woodland (below). Within three days, four new pairs had moved into the unoccupied areas (below, right) and some residents had expanded their territories. The new birds moved in from territories in hedgerows, considered to be suboptimal habitat. This type of territorial behaviour limits the density of breeding animals in areas of optimal habitat.

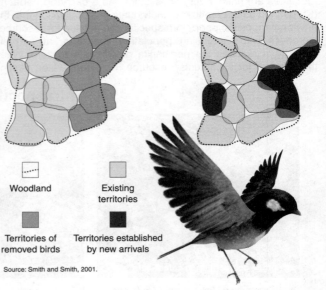

- ▱ Woodland
- ▨ Existing territories
- ▨ Territories of removed birds
- ■ Territories established by new arrivals

Source: Smith and Smith, 2001.

2. In the tank experiment with *Rana* (on the previous page), the tadpoles were contained in a fixed volume with a set amount of food:

(a) Describe how *Rana* tadpoles respond to resource limitation: _____

(b) Categorise the effect on the tadpoles as density-dependent / density-independent (delete one).

(c) Comment on how much the results of this experiment are likely to represent what happens in a natural population: _____

3. Identify two ways in which animals can reduce the intensity of intraspecific competition:

(a) _____

(b) _____

4. (a) Suggest why carrying capacity of an ecosystem might decline: _____

(b) Predict how a decline in carrying capacity might affect final population size: _____

5. Using appropriate examples, discuss the role of territoriality in reducing intraspecific competition: _____

Practical Ecology

Investigating populations in the field

Field studies: sampling populations and communities, measuring and interpreting the role of abiotic factors

Learning Objectives

☐ 1. Compile your own glossary from the **KEY WORDS** displayed in **bold type** in the learning objectives below.

Sampling Populations *(pages 74-76, 79-92, 95)*

☐ 2. Identify the type of information that can be gained from population studies (e.g. **abundance**, **density**, **age structure**, **distribution**). Explain why we **sample** populations and describe the advantages and drawbacks involved.

☐ 3. A field study should enable you to test a hypothesis about a certain aspect of a population. You should provide an outline of your study including reference to the type of data you will collect and the methods you will use, the size of your sampling unit (e.g. quadrat size) and the number of samples you will take, the assumptions of your investigation, and controls.

☐ 4. Explain how and why sample size affects the accuracy of population estimates. Explain how you would decide on a suitable sample size. Discuss the compromise between sampling accuracy and sampling effort.

☐ 5. Describe the following techniques used to study aspects of populations (e.g. **distribution**, **abundance**, **density**). Identify the advantages and limitations of each method with respect to sampling time, cost, and the suitability to the organism and specific habitat type:
   (a) **Direct counts**
   (b) **Frame** and/or **point quadrats**
   (c) **Belt** and/or **line transects**

   (d) **Mark and recapture** and the **Lincoln index**
   (e) **Netting** and **trapping**

☐ 6. Recognise the value to population studies of **radio-tracking** and **indirect methods** of sampling such as counting nests, and recording calls and droppings.

☐ 7. Describe **qualitative methods** for investigating the distribution of organisms in specific habitats.

☐ 8. Describe the methods used to ensure **random sampling**, and appreciate why this is important.

☐ 9. Recognise the appropriate ways in which different types of data may be recorded, analysed, and presented. Demonstrate an ability to apply **statistical tests**, as appropriate, to population data. Recognise that the type of data collected and the design of the study will determine how the data can be analysed.

☐ 10. Describe the role of **indicator organisms** in the assessment of water quality. Identify some common indicator organisms for freshwater systems and explain what their presence indicates.

Measuring Abiotic Factors *(pages 77-78, 93-94)*

☐ 11. Describe methods to measure abiotic factors in a habitat. Include reference to the following (as appropriate): pH, light, temperature, dissolved oxygen, current speed, total dissolved solids, and conductivity.

☐ 12. Appreciate the influence of abiotic factors on the distribution and abundance of organisms in a habitat.

See page 7 for additional details of these texts:

■ Adds, J., E. Larkcom, R. Miller, & R. Sutton, 1999. **Tools, Techniques and Assessment in Biology** (Thomas Nelson & sons Ltd.), as required.

■ Allen, D, M. Jones, and G. Williams, 2001. **Applied Ecology**, pp. 18-19, 26-33, 90-98

■ Cadogan, A. and M. Ingram, 2002 **Maths for Advanced Biology** (NelsonThornes), as required.

■ Reiss, M. & J. Chapman, 2000. **Environmental Biology** (Cambridge Uni. Press), pp. 6-14, 22-25.

■ Smith, R. L. & T.M. Smith, 2001. **Ecology and Field Biology**, reading as required.

Presentation MEDIA to support this topic:

ECOLOGY
• Practical Ecology

See page 7 for details of publishers of periodicals:

STUDENT'S REFERENCE

■ **Ecological Projects** Biol. Sci. Rev., 8(5) May 1996, pp. 24-26. *A thorough guide to planning and carrying out a field-based project.*

■ **Fieldwork - Sampling Animals** Biol. Sci. Rev., 10(4) March 1998, pp. 23-25. *Appropriate methodology for collecting animals in the field.*

■ **Fieldwork Sampling - Plants** Biol. Sci. Rev., 10(5) May 1998, pp. 6-8. *Excellent article covering the methodology for sampling plant communities.*

■ **Bird Ringing** Biol. Sci. Rev., 14(3) Feb. 2002, pp. 14-19. *Techniques used in investigating populations of highly mobile organisms: mark and recapture, ringing techniques, and application of diversity indices.*

■ **British Butterflies in Decline** Biol. Sci. Rev., 14(4) April 2002, pp. 10-13. *Documented changes in the distribution of British butterfly species. This account includes a description of the techniques used to monitor changes in population numbers.*

■ **Bowels of the Beasts** New Scientist, 22 August 1998, pp. 36-39. *Analyses of the faeces of animals can reveal much about the make-up, size, and genetic diversity of a population.*

TEACHER'S REFERENCE

■ **Ecology Fieldwork in 16 to 19 Biology** SSR, 84(307) December 2002, pp. 87. *Examines the fieldwork opportunities provided to 16-19 students and suggests how these could be enhanced and firmly established for all students studying biology.*

See pages 4-5 for details of how to access **Bio Links** from our web site: **www.thebiozone.com** From Bio Links, access sites under the topics:

ECOLOGY: > **Environmental Monitoring:** • Remote sensing and monitoring ... *and others* > **Populations and Communities:** • Communities • Quantitative population ecology • Sirtracking for wildlife research

Sampling Populations

Information about the populations of rare organisms in isolated populations may, in some instances, be collected by direct measure (direct counts and measurements of all the individuals in the population). However, in most cases, populations are too large to be examined directly and they must be sampled in a way that still provides information about them. Most practical exercises in population ecology involve the collection or census of living organisms, with a view to identifying the species and quantifying their abundance and other population features of interest. Sampling techniques must be appropriate to the community being studied and the information you wish to obtain. Some of the common strategies used in ecological sampling, and the situations for which they are best suited, are outlined in the table below. It provides an overview of points to consider when choosing a sampling regime. One must always consider the time and equipment available, the organisms involved, and the impact of the sampling method on the environment. For example, if the organisms involved are very mobile, sampling frames are not appropriate. If it is important not to disturb the organisms, observation alone must be used to gain information.

Method	Equipment and procedure	Information provided and considerations for use
Point sampling Random Systematic (grid)	Individual points are chosen on a map (using a grid reference or random numbers applied to a map grid) and the organisms are sampled at those points. Mobile organisms may be sampled using traps or nets.	**Useful for**: Determining species abundance and community composition. If samples are large enough, population characteristics (e.g. age structure, reproductive parameters) can be determined. **Considerations**: Time efficient. Suitable for most organisms. Depending on methods used, disturbance to the environment can be minimised. Species occurring in low abundance may be missed.
Transect sampling	Lines are drawn across a map and organisms occurring along the line are sampled. **Line transects**: Tape or rope marks the line. The species occurring on the line are recorded (all along the line or, more usually, at regular intervals). Lines can be chosen randomly (left) or may follow an environmental gradient. **Belt transects**: A measured strip is located across the study area to highlight any transitions. Quadrats are used to sample the plants and animals at regular intervals along the belt. Plants and immobile animals are easily recorded. Mobile or cryptic animals need to be trapped or recorded using appropriate methods.	**Useful for**: Well suited to determining changes in community composition along an environmental gradient. When placed randomly, they provide a quick measure of species occurrence. **Considerations for line transects**: Time efficient. Most suitable for plants and immobile or easily caught animals. Disturbance to the environment can be minimised. Species occurring in low abundance may be missed. **Considerations for belt transects**: Time consuming to do well. Most suitable for plants and immobile or easily caught animals. Good chance of recording most or all species. Efforts should be made to minimise disturbance to the environment.
Quadrat sampling	Sampling units or quadrats are placed randomly or in a grid pattern on the sample area. The occurrence of organisms in these squares is noted. Plants and slow moving animals are easily recorded. Rapidly moving or cryptic animals need to be trapped or recorded using appropriate methods.	**Useful for**: Well suited to determining community composition and features of population abundance: species density, frequency of occurrence, percentage cover, and biomass (if harvested). **Considerations**: Time consuming to do well. Most suitable for plants and immobile or easily caught animals. Quadrat size must be appropriate for the organisms being sampled and the information required. Some disturbance if organisms are removed.
Mark and recapture (capture-recapture) First sample: marked Second sample: proportion recaptured	Animals are captured, marked, and then released. After a suitable time period, the population is resampled. The number of marked animals recaptured in a second sample is recorded as a proportion of the total.	**Useful for**: Determining total population density for highly mobile species in a certain area (e.g. butterflies). Movements of individuals in the population can be tracked (especially when used in conjunction with electronic tracking devices). **Considerations**: Time consuming to do well. Not suitable for immobile species. Population should have a finite boundary. Period between samplings must allow for redistribution of marked animals in the population. Marking should present little disturbance and should not affect behaviour.

In the transect sampling illustration: 0.5 m, with label "Environmental gradient".

1. Explain why we **sample** populations: _____

   _____

2. Describe a sampling technique that would be appropriate for determining each of the following:

   (a) The percentage cover of a plant species in pasture: _____

   (b) The density and age structure of a plankton population: _____

   (c) Change in community composition from low to high altitude on a mountain: _____

Designing Your Field Study

The following provides an example and some ideas for designing a field study. It provides a framework which can be modified for most simple comparative field investigations. For reasons of space, the full methodology is not included.

Pill millipede
Glomeris marginata

Oak woodland

Coniferous woodland

Observation

A student read that a particular species of pill millipede (left) is extremely abundant in forest leaf litter, but a search in the litter of a conifer-dominated woodland near his home revealed only very low numbers of this millipede species.

Hypothesis

This millipede species is adapted to a niche in the leaf litter of oak woodlands and is abundant there. However, it is rare in the litter of coniferous woodland. The **null hypothesis** is that there is no difference between the abundance of this millipede species in oak and coniferous woodland litter.

Oak or coniferous woodland

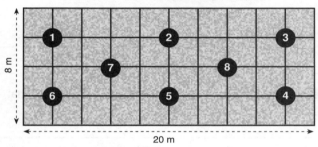

8 m

20 m

 Sampling sites numbered 1-8 at evenly spaced intervals on a 2 x 2 m grid within an area of 20 m x 8 m.

Sampling Programme

A sampling programme was designed to test the prediction that the millipedes are more abundant in the leaf litter of oak woodlands than in coniferous woodlands.

Equipment and Procedure

Sites: For each of the two woodland types, an area 20 x 8 m was chosen and marked out in 2 x 2 m grids. Eight sampling sites were selected, evenly spaced along the grid as shown.

- The general area for the study chosen was selected on the basis of the large amounts of leaf litter present.
- Eight sites were chosen as the largest number feasible to collect and analyse in the time available.
- The two woodlands were sampled on sequential days.

Capture of millipedes: At each site, a 0.4 x 0.4 m quadrat was placed on the forest floor and the leaf litter within the quadrat was collected. Millipedes and other leaf litter invertebrates were captured using a simple gauze lined funnel containing the leaf litter from within the quadrat. A lamp was positioned over each funnel for two hours and the invertebrates in the litter moved down and were trapped in the collecting jar.

- After two hours each jar was labelled with the site number and returned to the lab for analysis.
- The litter in each funnel was bagged, labelled with the site number and returned to the lab for weighing.
- The number of millipedes at each site was recorded.
- The numbers of other invertebrates (classified into major taxa) were also noted for reference.

Sampling equipment: leaf litter light trap

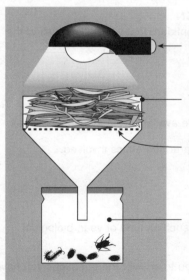

Light from a battery operated lamp drives the invertebrates down through the leaf litter.

Large (diameter 300 mm) funnel containing leaf litter resting on a gauze platform.

Gauze allows invertebrates of a certain size to move down the funnel.

Collecting jar placed in the litter on the forest floor traps the invertebrates that fall through the gauze and prevents their escape.

Assumptions

- The areas chosen in each woodland were representative of the woodland types in terms of millipede abundance.
- Eight sites were sufficient to adequately sample the millipede populations in each forest.
- A quadrat size of 0.4 x 0.4 m contained enough leaf litter to adequately sample the millipedes at each site.
- The millipedes were not preyed on by any of the other invertebrates captured in the collecting jar.
- All the invertebrates within the quadrat were captured.
- Millipedes moving away from the light are effectively captured by the funnel apparatus and cannot escape.
- Two hours was long enough for the millipedes to move down through the litter and fall into the trap.

Note that these last two assumptions could be tested by examining the bagged leaf litter for millipedes after returning to the lab.

Notes on collection and analysis
- Mean millipede abundance was calculated from the counts from the eight sites. The difference in abundance at the sites was tested using a Student's *t* test.
- After counting and analysis of the samples, all the collected invertebrates were returned to the sites.

Practical Ecology

A note about sample size

When designing a field study, the size of your sampling unit (e.g. quadrat size) and the sample size (the number of samples you will take) should be major considerations. There are various ways to determine the best quadrat size. Usually, these involve increasing the quadrat size until you stop finding new species. For simple field studies, the number of samples you take (the sample size or *n* value) will be determined largely by the resources and time that you have available to collect and analyse your data. It is usually best to take as many samples as you can, as this helps to account for any natural variability present and will give you greater confidence in your data. These aspects of study design as well as coverage of collecting methods are covered in many field ecology manuals and texts covering statistics for biology.

1. Explain the importance of recognising any assumptions that you are making in your study:

   _____

   _____

2. Describe how you would test whether the quadrat size of 0.4 x 0.4 m was adequate to effectively sample the millipedes:

   _____

   _____

3. Suggest why the litter was bagged, returned to the lab and then weighed properly for the analysis:

   _____

   _____

4. Suggest why the numbers of other invertebrates were also recorded even though it was only millipede abundance that was being investigated:

   _____

   _____

   _____

   _____

YOUR CHECKLIST FOR FIELD STUDY DESIGN

The following provides a checklist for a field study. Check off the points when you are confident that you have satisfied the requirements in each case:

1. **Preliminary:**

   ☐ (a) Makes a hypothesis based on observation(s).

   ☐ (b) The hypothesis (and its predictions) are testable using the resources you have available (the study is feasible).

   ☐ (c) The organism you have chosen is suitable for the study and you have considered the ethics involved.

2. **Assumptions and site selection:**

   ☐ (a) You are aware of any assumptions that you are making in your study.

   ☐ (b) You have identified aspects of your field design that could present problems (such as time of year, biological rhythms of your test organism, difficulty in identifying suitable habitats etc.).

   ☐ (c) The study sites you have selected have the features necessary in order for you to answer the questions you have asked in your hypothesis.

3. **Data collection:**

   ☐ (a) You are happy with the way in which you are going to take your measurements or samples.

   ☐ (b) You have considered the size of your sampling unit and the number of samples you are going to take (and tested for these if necessary).

   ☐ (c) You have given consideration to how you will analyse the data you collect and made sure that your study design allows you to answer the questions you wish to answer.

Monitoring Physical Factors

Most ecological studies require us to measure the physical factors (parameters) in the environment that may influence the abundance and distribution of organisms. In recent years there have been substantial advances in the development of portable, light-weight meters and dataloggers. These enable easy collection and storage of data in the field.

Quantum light meter: Measures light intensity levels. It is not capable of measuring light quality (wavelength).

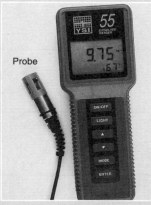

Dissolved oxygen meter: Measures the amount of oxygen dissolved in water (expressed as mgl⁻¹).

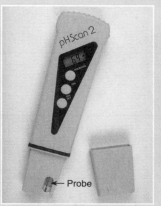

pH meter: Measures the acidity of water or soil, if it is first dissolved in pure water (pH scale 0 to 14).

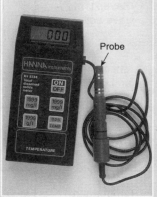

Total dissolved solids (TDS) meter: Measures content of dissolved solids (as ions) in water in mgl⁻¹.

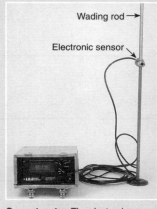

Current meter: The electronic sensor is positioned at set depths in a stream or river on the calibrated wading rod as current readings are taken.

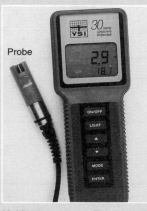

Multipurpose meter: This is a multi-functional meter, which can measure salinity, conductivity and temperature simply by pushing the MODE button.

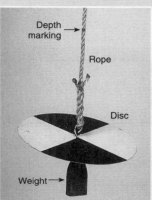

Secchi disc: This simple device is used to provide a crude measure of water clarity (the maximum depth at which the disc can just be seen).

Collecting a water sample: A Nansen bottle is used to collect water samples from a lake for lab analysis, testing for nutrients, oxygen and pH.

Practical Ecology

Dataloggers and Environmental Sensors

Dataloggers are electronic instruments that record measurements over time. They are equipped with a microprocessor, data storage facility, and sensor. Different sensors are employed to measure a range of variables in water (photos A and B) or air (photos C and D), as well as make physiological measurements. The datalogger is connected to a computer, and software is used to set the limits of operation (e.g. the sampling interval) and initiate the logger. The logger is then disconnected and used remotely to record and store data. When reconnected to the computer, the data are downloaded, viewed, and plotted. Dataloggers, such as those pictured here from PASCO, are being increasingly used in professional and school research. They make data collection quick and accurate, and they enable prompt data analysis.

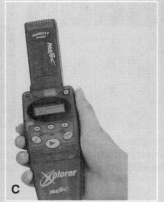

Dataloggers are now widely used to monitor conditions in aquatic environments. Different variables such as pH, temperature, conductivity, and dissolved oxygen can be measured by changing the sensor attached to the logger.

Dataloggers fitted with sensors are portable and easy to use in a wide range of terrestrial environments. They are used to measure variables such as air temperature and pressure, relative humidity, light, and carbon dioxide gas.

Code: RA 2

1. The physical factors of an exposed rocky shore and a sheltered estuarine mudflat differ markedly. For each of the factors listed in the table below, briefly describe how they may differ (if at all):

Environmental parameter	Exposed rocky coastline	Estuarine mudflat
Severity of wave action		
Light intensity and quality		
Salinity/ conductivity		
Temperature change (diurnal)		
Substrate/ sediment type		
Oxygen concentration		
Exposure time to air (tide out)		

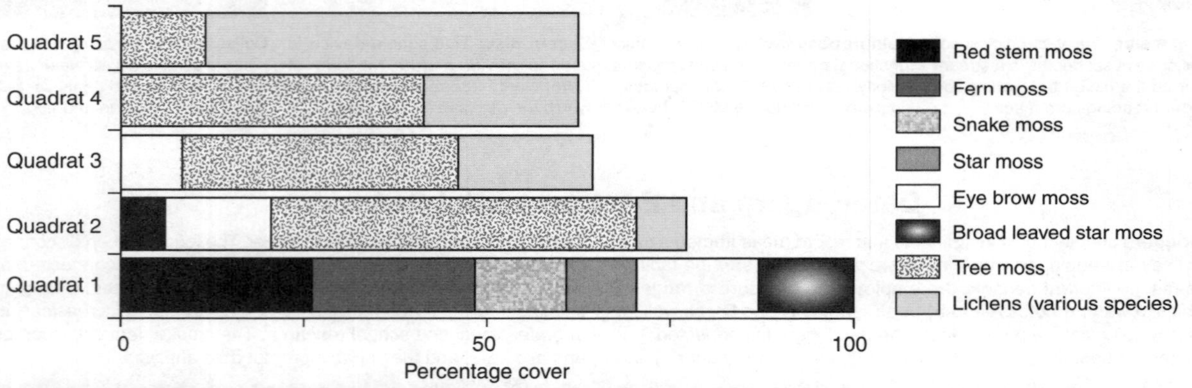

Legend:
- Red stem moss
- Fern moss
- Snake moss
- Star moss
- Eye brow moss
- Broad leaved star moss
- Tree moss
- Lichens (various species)

Quadrats 1–5 shown, Percentage cover axis 0, 50, 100.

QUADRAT	1	2	3	4	5
Height (m)	0.4	0.8	1.2	1.6	2.0
Light (arbitrary units)	40	56	68	72	72
Humidity (percent)	99	88	80	76	78
Temperature (°C)	12.1	12.2	13	14.3	14.2

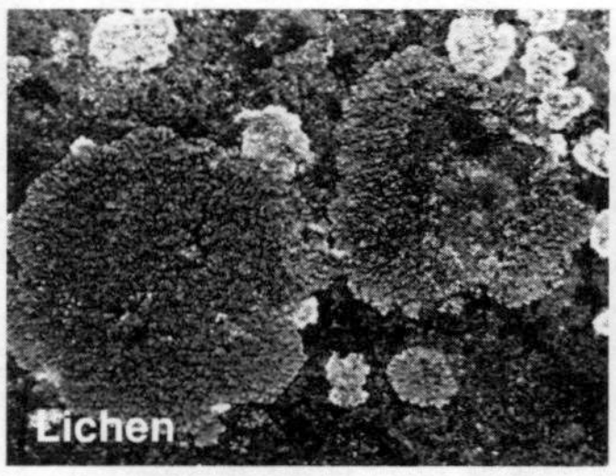

Lichen

Moss

2. The figure (above) shows the changes in vegetation cover along a 2 m vertical transect up the trunk of an oak tree (*Quercus*). Changes in the physical factors light, humidity, and temperature along the same transect were also recorded. From what you know about the ecology of mosses and lichens, account for the observed vegetation distribution:

_____

_____

_____

Indirect Sampling

If populations are small and easily recognised they may be monitored by direct measurement quite easily. However, direct measurement of elusive, or widely dispersed populations is not always feasible. In these cases, indirect methods can be used to assess population abundance, provide information on habitat use and range, and enable biologists to link habitat quality to species presence or absence. Indirect sampling methods provide less reliable measures of abundance than direct sampling, but are widely used nevertheless. They rely on recording the signs of a species, e.g. scat, calls, tracks, and markings on vegetation, and

using these to assess population abundance. In Australia, the Environmental Protection Agency (EPA) provides a Frog Census Datasheet (below) on which volunteers record details about frog populations and habitat quality in their area. This programme enables the EPA to gather information across Australia. Another example of indirect population sampling in New Zealand is the Kiwi Recovery Programme (see following page). Kiwi are elusive, nocturnal birds and all species are endangered. Conservationists rely on public recordings of kiwi to monitor population distribution and target their conservation effort appropriately.

INFORMATION NEEDED FOR THE FROG CENSUS

- Where you recorded frogs calling; When you made the recordings; and What frogs you recorded (if possible).

Observers Name:_____
Contact Address: _____

Post Code: _____
Telephone Home: _____   Work / Mobile:_____

Do You Want to be involved next year?(Please Circle)

Location Description (Try to provide enough detail to enable us to find map.
Please use a separate datasheet for each site)

is location the same as in (CIRCLE)   1994   1995   1996   1997

Grid Reference of Location and Type of Map Used: _____
OR Street Directory Reference: Year and Edition: __
Page Number: _____   Grid Reference: _____
Nearest Town from Location (if known): _____

Date of Observation (e.g. 8 Sept 1998): _____
Time Range of Observation (e.g. 8.30-8.40 pm): _____

HABITAT ASSESSMENT

Habitat Type (please circle one):   pond   dam   stream   drain
                          reservoir   wetland   spring   swamp
Comments:_____

WATER QUALITY and WEATHER

CIRCLE to indicate the condition of the site (you can circle more than one choice).

Water Flow: Still   Flowing Slowly Flowing Quickly
Water Appearance: Clear   Polluted   Frothy Oily
                     Muddy
Weather Conditions: 1. Windy / Still
                     2. Overcast / Recent Rains / Dry (indicate for 1 AND 2)

FROGS HEARD CALLING

Please indicate your estimate of how many frogs you heard calling (NOTE it is very important to tell us if you heard no frogs)
Number of Calls Heard (circle):
None   One   Few (2-9)   Many (10-50)   Lots (>50)

If you want to test your frog knowledge write the species you heard calling:
Species of Frog(s) Identified: 1._____   2._____
                               3._____   4._____
Comments:_____

Now we need you to return your datasheet and tape in the postage free post-pak addressed to REPLY PAID 6350 Mr Peter Goonan Environment Protection Agency GPO Box 2607 ADELAIDE SA 5001. We will identify your frog calls and let you the results of your recordings.

Office use only. Please leave blank.
FROG SPECIES PRESENT.

...ies Number	Species 1	Species 2	Species 3	Species 4	Species 5
...ies Name					
(2 - 9)					
y (10 - 50)					
>50)					

ENVIRONMENT PROTECTION AGENCY
DEPARTMENT FOR ENVIRONMENT HERITAGE AND ABORIGINAL AFFAIRS

Recording a date and accurate map reference is important

Population estimates are based on the number of frog calls recorded by the observer

For many elusive, nocturnal species, burrows and probe holes can be used to provide a rather crude assessment of population presence in an area. Experienced biologists may be able to tell whether or not the burrow is still in use. This photograph shows a **spider burrow**; its distinctive appearance identifies the species that made it.

Gull tracks

The analysis of animal tracks allows wildlife biologists to identify habitats in which animals live and to conduct population surveys. Interpreting tracks accurately requires considerable skill as tracks may vary in appearance even when from the same individual. Tracks are particularly useful as a way to determine habitat use and preference.

Deer droppings

All animals leave scats (faeces) which are species specific and readily identifiable. Scats can be a valuable tool by which to gather data from elusive, nocturnal, easily disturbed, or highly mobile species. Faecal analyses can provide information on diet, movements, population density, sex ratios, age structure, and even genetic diversity.

Practical Ecology

1. Describe two kinds of indirect signs that could be used to detect the presence of frogs:

   (a) _____   (b) _____

2. (a) Describe the kind of information that the EPA would gather from their Frog Census Datasheet: _____

   _____

   (b) Explain a use for this information: _____

   _____

   _____

Code: RA 1

The **Kiwi Reporting Card** is issued to trampers and conservation groups who are helping New Zealand's Department of Conservation to gather census data on kiwi. Read all parts of the card carefully and answer the questions below.

DATE:	OBSERVER'S DETAILS:	**KIWI REPORTING CARD**

LOCATION:	NAME	Please complete this form as fully as possible and return to the Department of Conservation. **KIWI RECOVERY**
	ADDRESS	

MAP SERIES	SHEET	GRID REFERENCE	PHONE Home/Work	**NOTES**

GRID REFERENCE: ☐☐☐ E ☐☐☐ N

NUMBER OF KIWI SEEN (Other, please specify)

COMMENTS (Vegetation, habitat, dogs or other predators seen)

NOTES
1 The call of the male kiwi is a repetitive (8-25 notes) high-pitched whistle.
2 The call of the female kiwi is a repetitive (10-20 notes) coarse rasping note.
3 Weka, moreporks and possums are often confused with kiwi calls.
4 Footprints are about the size of a domestic chicken and are often found in mud or snow.
5 Probeholes usually occur in groups and look like a screwdriver has been pushed into the ground, rotated and pulled out again. They are about 10cm deep.
6 This form may also be used to report other species of wildlife such as kaka, kokako, blue duck, bats, etc. Please ensure you clearly identify what you are recording.
7 Post this card to the Department of Conservation (address provided on the back ...... it with a hut warden.

NUMBER OF KIWI HEARD		SIGNS OF KIWI PRESENT (eg. footprints, probeholes)
Male calls	Female calls	

Are you 100% sure that what you saw or heard was a kiwi? YES/NO

3. Describe three kinds of indirect signs that can be used to detect the presence of kiwi:

(a) _____

(b) _____

(c) _____

4. Explain why it is not easy to carry out a direct count of a kiwi population:

_____

_____

_____

5. Describe the attempts that the organisers of this data collecting programme have made to ensure that the people recording their observations are correctly identifying kiwi signs:

_____

_____

_____

6. Explain why the comments section on the card requests information on the habitat, dogs, or other predators seen:

_____

_____

_____

7. Using an example, describe another indirect method of population sampling and outline its advantages and drawbacks:

_____

_____

_____

_____

_____

_____

Quadrat Sampling

Quadrat sampling is a method by which organisms in a certain proportion (sample) of the habitat are counted directly. As with all sampling methods, it is used to estimate population parameters when the organisms present are too numerous to count in total. It can be used to estimate population **abundance** (number), **density**, **frequency of occurrence**, and **distribution**. Quadrats may be used without a transect when studying a relatively uniform habitat. In this case, the quadrat positions are chosen randomly using a random number table.

The general procedure is to count all the individuals (or estimate their percentage cover) in a number of quadrats of known size and to use this information to work out the abundance or percentage cover value for the whole area. The number of quadrats used and their size should be appropriate to the type of organism involved (e.g. grass vs tree).

$$\text{Estimated average density} = \frac{\text{Total number of individuals counted}}{\text{Number of quadrats} \times \text{area of each quadrat}}$$

Guidelines for Quadrat Use:

1. The **area of each quadrat** must be known exactly and ideally quadrats should be the same shape. The quadrat does not have to be square (it may be rectangular, hexagonal etc.).

2. **Enough quadrat samples** must be taken to provide results that are representative of the total population.

3. The **population of each quadrat** must be known exactly. Species must be distinguishable from each other, even if they have to be identified at a later date. It has to be decided beforehand what the count procedure will be and how organisms over the quadrat boundary will be counted.

4. The size of the quadrat should be appropriate to the organisms and habitat, e.g. a large size quadrat for trees.

5. The quadrats must be **representative of the whole area**. This is usually achieved by **random sampling** (right).

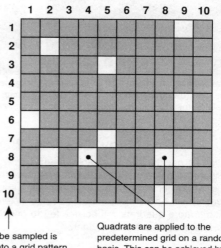

The area to be sampled is divided up into a grid pattern with indexed coordinates

Quadrats are applied to the predetermined grid on a random basis. This can be achieved by using a random number table.

Sampling a centipede population

A researcher by the name of Lloyd (1967) sampled centipedes in Wytham Woods, near Oxford in England. A total of 37 hexagon–shaped quadrats were used, each with a diameter of 30 cm (see diagram on right). These were arranged in a pattern so that they were all touching each other. Use the data in the diagram to answer the following questions.

1. Determine the average number of centipedes captured per quadrat:

   _____

2. Calculate the estimated average density of centipedes per square metre (remember that each quadrat is 0.08 square metres in area):

   _____

3. Looking at the data for individual quadrats, describe in general terms the distribution of the centipedes in the sample area:

   _____

   _____

4. Describe one factor that might account for the distribution pattern:

   _____

   _____

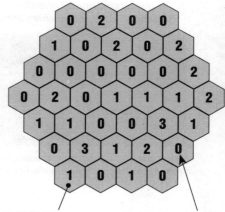

Each quadrat was a hexagon with a diameter of 30 cm and an area of 0.08 square metres.

The number in each hexagon indicates how many centipedes were caught in that quadrat.

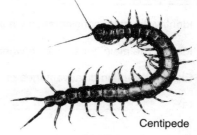

Centipede

Code: DA 2

Quadrat-Based Estimates

The simplest description of a plant community in a habitat is a list of the species that are present. This qualitative assessment of the community has the limitation of not providing any information about the **relative abundance** of the species present. Quick estimates can be made using **abundance scales**, such as the ACFOR scale described below. Estimates of percentage cover provide similar information. These methods require the use of **quadrats**. Quadrats are used extensively in plant ecology. This activity outlines some of the common considerations when using quadrats to sample plant communities.

What Size Quadrat?

Quadrats are usually square, and cover 0.25 m^2 (0.5 m x 0.5 m) or 1 m^2, but they can be of any size or shape, even a single point. The quadrats used to sample plant communities are often 0.25 m^2. This size is ideal for low-growing vegetation, but quadrat size needs to be adjusted to habitat type. The quadrat must be large enough to be representative of the community, but not so large as to take a very long time to use.

A quadrat covering an area of 0.25 m^2 is suitable for most low growing plant communities, such as this alpine meadow, fields, and grasslands.

Larger quadrats (e.g. 1 m^2) are needed for communities with shrubs and trees. Quadrats as large as 4 m x 4 m may be needed in woodlands.

Small quadrats (0.01 m^2 or 100 mm x 100 mm) are appropriate for lichens and mosses on rock faces and tree trunks.

How Many Quadrats?

As well as deciding on a suitable quadrat size, the other consideration is how many quadrats to take (the sample size). In species-poor or very homogeneous habitats, a small number of quadrats will be sufficient. In species-rich or heterogeneous habitats, more quadrats will be needed to ensure that all species are represented adequately.

Determining the number of quadrats needed

- Plot the cumulative number of species recorded (on the *y* axis) against the number of quadrats already taken (on the *x* axis).
- The point at which the curve levels off indicates the suitable number of quadrats required.

Fewer quadrats are needed in species-poor or very uniform habitats, such as this bluebell woodland.

Describing Vegetation

Density (number of individuals per unit area) is a useful measure of abundance for animal populations, but can be problematic in plant communities where it can be difficult to determine where one plant ends and another begins. For this reason, plant abundance is often assessed using **percentage cover**. Here, the percentage of each quadrat covered by each species is recorded, either as a numerical value or using an abundance scale such as the ACFOR scale.

The ACFOR Abundance Scale

A = Abundant (30% +)
C = Common (20-29%)
F = Frequent (10-19%)
O = Occasional (5-9%)
R = Rare (1-4%)

The AFCOR scale could be used to assess the abundance of species in this wildflower meadow. Abundance scales are subjective, but it is not difficult to determine which abundance category each species falls into.

1. Describe one difference between the methods used to assess species abundance in plant and in animal communities:

   _____

2. Identify the main consideration when determining appropriate quadrat size: _____

3. Identify the main consideration when determining number of quadrats: _____

4. Explain two main disadvantages of using the ACFOR abundance scale to record information about a plant community:

   (a) _____

   (b) _____

Sampling a Leaf Litter Population

The diagram on the following page represents an area of leaf litter from a forest floor with a resident population of organisms. The distribution of four animal species as well as the arrangement of leaf litter is illustrated. Leaf litter comprises leaves and debris that have dropped off trees to form a layer of detritus. This exercise is designed to practice the steps required in planning and carrying out a sampling of a natural population. It is desirable, but not essential, that students work in groups of 2–4.

1. Decide on the sampling method

For the purpose of this exercise, it has been decided that the populations to be investigated are too large to be counted directly and a quadrat sampling method is to be used to estimate the average density of the four animal species as well as that of the leaf litter.

2. Mark out a grid pattern

Use a ruler to mark out 3 cm intervals along each side of the sampling area (area of quadrat = 0.03 x 0.03 m). **Draw lines** between these marks to create a 6 x 6 grid pattern (total area = 0.18 x 0.18 m). This will provide a total of 36 quadrats that can be investigated.

3. Number the axes of the grid

Only a small proportion of the possible quadrat positions are going to be sampled. It is necessary to select the quadrats in a random manner. It is not sufficient to simply guess or choose your own on a 'gut feeling'. The best way to choose the quadrats randomly is to create a numbering system for the grid pattern and then select the quadrats from a random number table. Starting at the *top left hand corner*, **number the columns** and **rows** from 1 to 6 on each axis.

4. Choose quadrats randomly

To select the required number of quadrats randomly, use random numbers from a random number table. The random numbers are used as an index to the grid coordinates. Choose 6 quadrats from the total of 36 using table of random numbers provided for you at the bottom of the facing page. Make a note of which column of random numbers you choose. Each member of your group should choose a different set of random numbers (i.e. different column: A–D) so that you can compare the effectiveness of the sampling method.

Column of random numbers chosen: _____

NOTE: Highlight the boundary of each selected quadrat with coloured pen/highlighter.

5. Decide on the counting criteria

Before the counting of the individuals for each species is carried out, the criteria for counting need to be established.

There may be some problems here. You must decide before sampling begins as to what to do about individuals that are only partly inside the quadrat. Possible answers include:

(a) Only counting individuals if they are completely inside the quadrat.

(b) Only counting individuals that have a clearly defined part of their body inside the quadrat (such as the head).

(c) Allowing for 'half individuals' in the data (e.g. 3.5 snails).

(d) Counting an individual that is inside the quadrat by half or more as one complete individual.

Discuss the merits and problems of the suggestions above with other members of the class (or group). You may even have counting criteria of your own. Think about other factors that could cause problems with your counting.

6. Carry out the sampling

Carefully examine each selected quadrat and **count the number of individuals** of each species present. Record your data in the spaces provided on the facing page.

7. Calculate the population density

Use the combined data TOTALS for the sampled quadrats to estimate the average density for each species by using the formula:

Density =

$$\frac{\text{Total number in all quadrats sampled}}{\text{Number of quadrats sampled} \ \ \textbf{X} \ \ \text{area of a quadrat}}$$

Remember that a total of 6 quadrats are sampled and each has an area of 0.0009 m^2. The density should be expressed as the number of individuals *per square metre (no. m^{-2})*.

Woodlouse: [   ]   False scorpion: [   ]

Centipede: [   ]   Leaf: [   ]

Springtail: [   ]

8. (a) In this example the animals are not moving. Describe the problems associated with sampling moving organisms. Explain how you would cope with sampling these same animals if they were really alive and very active:

_____

_____

_____

(b) Carry out a direct count of all 4 animal species and the leaf litter for the whole sample area (all 36 quadrats). Apply the data from your direct count to the equation given in (7) above to calculate the actual population density (remember that the number of quadrats in this case = 36):

Woodlouse: [   ]   Centipede: [   ]   False scorpion: [   ]   Springtail: [   ]   Leaf: [   ]

Compare your estimated population density to the actual population density for each species:

_____

_____

_____

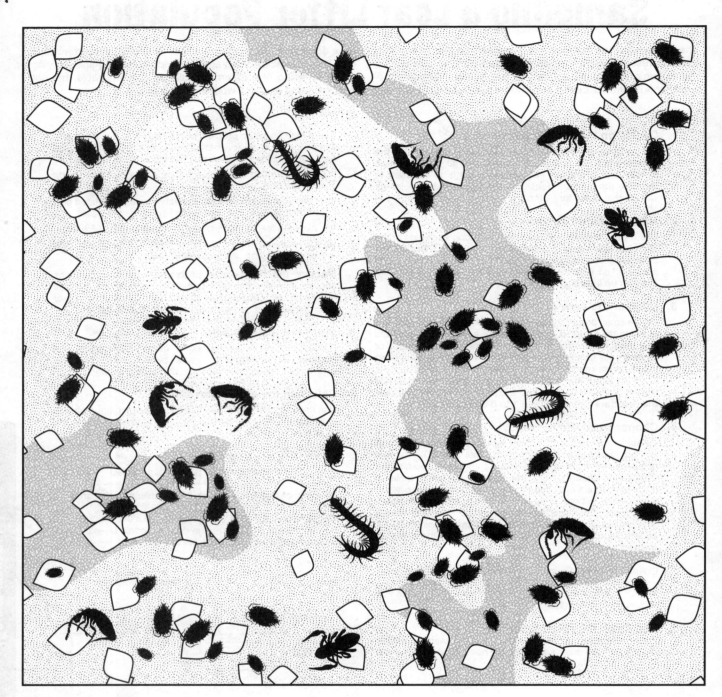

Coordinates for each quadrat	Woodlouse	Centipede	False scorpion	Springtail	Leaf
1:					
2:					
3:					
4:					
5:					
6:					
TOTAL					

Table of random numbers

A	B	C	D
2 2	3 1	6 2	2 2
3 2	1 5	6 3	4 3
3 1	5 6	3 6	6 4
4 6	3 6	1 3	4 5
4 3	4 2	4 5	3 5
5 6	1 4	3 1	1 4

The table above has been adapted from a table of random numbers from a statistics book. Use this table to select quadrats randomly from the grid above. Choose one of the columns (A to D) and use the numbers in that column as an index to the grid. The first digit refers to the row number and the second digit refers to the column number. To locate each of the 6 quadrats, find where the row and column intersect, as shown below:

Example: **5 2** refers to the 5th row and the 2nd column

Transect Sampling

A **transect** is a line placed across a community of organisms. Transects are usually carried out to provide information on the **distribution** of species in the community. This is of particular value in situations where environmental factors change over the sampled distance. This change is called an **environmental gradient** (e.g. up a mountain or across a seashore). The usual practice for small transects is to stretch a string between two markers. The string is marked off in measured distance intervals, and the species at each marked point are noted. The sampling points along the transect may also be used for the siting of quadrats, so that changes in density and community composition can be recorded. Belt transects are essentially a form of continuous quadrat sampling. They provide more information on community composition but can be difficult to carry out. Some transects provide information on the vertical, as well as horizontal, distribution of species (e.g. tree canopies in a forest).

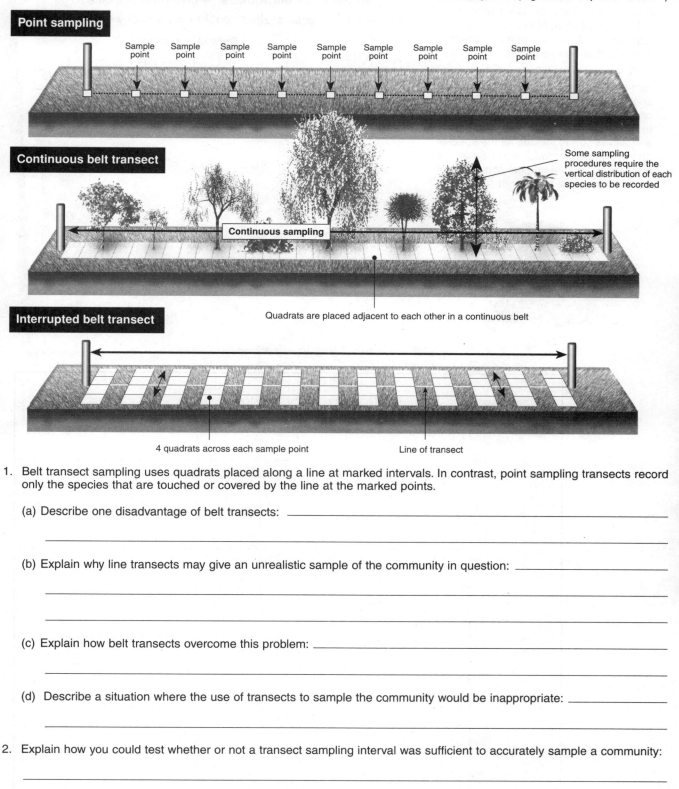

Point sampling

Sample point (×9)

Continuous belt transect

Some sampling procedures require the vertical distribution of each species to be recorded

Continuous sampling

Quadrats are placed adjacent to each other in a continuous belt

Interrupted belt transect

4 quadrats across each sample point

Line of transect

1. Belt transect sampling uses quadrats placed along a line at marked intervals. In contrast, point sampling transects record only the species that are touched or covered by the line at the marked points.

   (a) Describe one disadvantage of belt transects: _____

   _____

   (b) Explain why line transects may give an unrealistic sample of the community in question: _____

   _____

   _____

   (c) Explain how belt transects overcome this problem: _____

   _____

   (d) Describe a situation where the use of transects to sample the community would be inappropriate: _____

   _____

2. Explain how you could test whether or not a transect sampling interval was sufficient to accurately sample a community:

   _____

   _____

   _____

Code: DA 2

Kite graphs are an ideal way in which to present distributional data from a belt transect (e.g. abundance or percentage cover along an environmental gradient. Usually, they involve plots for more than one species. This makes them good for highlighting probable differences in habitat preference between species. Kite graphs may also be used to show changes in distribution with time (e.g. with daily or seasonal cycles).

3. The data on the right were collected from a rocky shore field trip. Periwinkles from four common species of the genus *Littorina* were sampled in a continuous belt transect from the low water mark, to a height of 10 m above that level. The number of each of the four species in a 1 m² quadrat was recorded.

Plot a **kite graph** of the data for all four species on the grid below. Be sure to choose a scale that takes account of the maximum number found at any one point and allows you to include all the species on the one plot. Include the scale on the diagram so that the number at each point on the kite can be calculated.

Field data notebook
Numbers of periwinkles (4 common species) showing vertical distribution on a rocky shore

Height above low water (m)	Periwinkle species: L. littorea	L. saxatalis	L. neritoides	L. littoralis
0-1	0	0	0	0
1-2	1	0	0	3
2-3	3	0	0	17
3-4	9	3	0	12
4-5	15	12	0	1
5-6	5	24	0	0
6-7	2	9	2	0
7-8	0	2	11	0
8-9	0	0	47	0
9-10	0	0	59	0

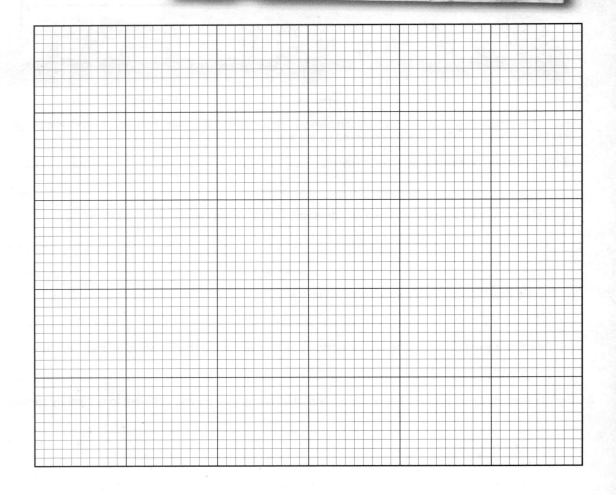

Mark and Recapture Sampling

The mark and recapture method of estimating population size is used in the study of animal populations where individuals are highly mobile. It is of no value where animals do not move or move very little. The number of animals caught in each sample must be large enough to be valid. The technique is outlined in the diagram below.

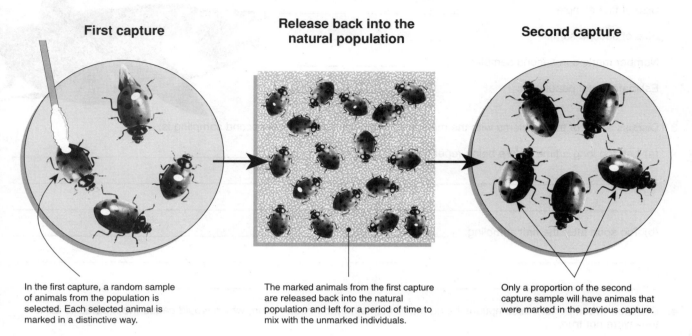

First capture

In the first capture, a random sample of animals from the population is selected. Each selected animal is marked in a distinctive way.

Release back into the natural population

The marked animals from the first capture are released back into the natural population and left for a period of time to mix with the unmarked individuals.

Second capture

Only a proportion of the second capture sample will have animals that were marked in the previous capture.

The Lincoln Index

$$\text{Total population} = \frac{\text{No. of animals in 1st sample (all marked)} \quad \text{X} \quad \text{Total no. of animals in 2nd sample}}{\text{Number of marked animals in the second sample (recaptured)}}$$

The mark and recapture technique comprises a number of simple steps:

1. The population is sampled by capturing as many of the individuals as possible and practical.

2. Each animal is marked in a way to distinguish it from unmarked animals (unique mark for each individual not required).

3. Return the animals to their habitat and leave them for a long enough period for complete mixing with the rest of the population to take place.

4. Take another sample of the population (this does not need to be the same sample size as the first sample, but it does have to be large enough to be valid).

5. Determine the numbers of marked to unmarked animals in this second sample. Use the equation above to estimate the size of the overall population.

Practical Ecology

1. For this exercise you will need several boxes of matches and a pen. Work in a group of 2-3 students to 'sample' the population of matches in the full box by using the mark and recapture method. Each match will represent one animal.

(a) Take out 10 matches from the box and mark them on 4 sides with a pen so that you will be able to recognise them from the other unmarked matches later.

(b) Return the marked matches to the box and shake the box to mix the matches.

(c) Take a sample of 20 matches from the same box and record the number of marked matches and unmarked matches.

(d) Determine the total population size by using the equation above.

(e) Repeat the sampling 4 more times (steps b–d above) and record your results:

	Sample 1	Sample 2	Sample 3	Sample 4	Sample 5
Estimated population					

(f) Count the actual number of matches in the matchbox : _____

(g) Compare the actual number to your estimates. By how much does it differ: _____

Code: PDA 2

2. In 1919 a researcher by the name of Dahl wanted to estimate the number of trout in a Norwegian lake. The trout were subject to fishing so it was important to know how big the population was in order to manage the fish stock. He captured and marked 109 trout in his first sample. A few days later, he caught 177 trout in his second sample, of which 57 were marked. Use the Lincoln index (on the previous page) to estimate the total population size:

Size of first sample: _____

Size of second sample: _____

Number marked in second sample: _____

Estimated total population: _____

3. Discuss some of the problems with the mark and recapture method if the second sampling is:

   (a) Left too long a time before being repeated: _____

   _____

   _____

   (b) Too soon after the first sampling: _____

   _____

   _____

4. Describe two important assumptions being made in this method of sampling, which would cause the method to fail if they were not true:

   (a) _____

   _____

   _____

   (b) _____

   _____

   _____

5. Some types of animal would be unsuitable for this method of population estimation (i.e. would not work).

   (a) Name an animal for which this method of sampling would not be effective: _____

   (b) Explain your answer above: _____

   _____

   _____

6. Describe three methods for marking animals for mark and recapture sampling. Take into account the possibility of animals shedding their skin, or being difficult to get close to again:

   (a) _____

   (b) _____

   (c) _____

7. At various times since the 1950s, scientists in the UK and Canada have been involved in computerised tagging programmes for Northern cod (a species once abundant in Northern Hemisphere waters but now severely depleted). Describe the type of information that could be obtained through such tagging programmes:

   _____

   _____

   _____

   _____

Sampling Animal Populations

Unlike plants, most animals are highly mobile and present special challenges in terms of sampling them **quantitatively** to estimate their distribution and abundance. The equipment available for sampling animals ranges from various types of nets and traps (below), to more complex electronic devices, such as those used for radio-tracking large mobile species.

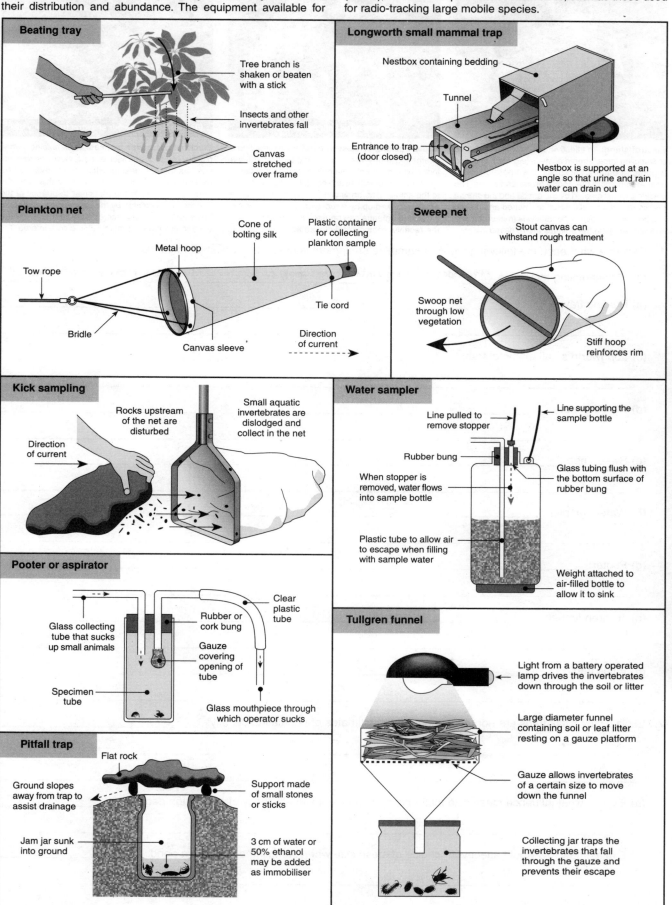

Practical Ecology

Code: RA 2

Electrofishing a stream (Sweden)

Tuatara fitted with transmitter

Electronic bat detector showing frequency dial

Electrofishing: An effective, but expensive method of sampling larger stream animals (e.g. fish). Wearing a portable battery backpack, the operator walks upstream holding the anode probe and a net. The electrical circuit created by the anode and the stream bed stuns the animals, which are netted and placed in a bucket to recover. After analysis (measurement, species, weights) the animals are released.

Radio-tracking: A relatively non-invasive method of examining many features of animal populations, including movement, distribution, and habitat use. A small transmitter with an antenna (arrowed) is attached to the animal. The transmitter emits a pulsed signal which is picked up by a receiver. In difficult terrain, a tracking antenna can be used in conjunction with the receiver to accurately fix an animal's position.

Electronic detection devices: To sample nocturnal, highly mobile species such as bats, electronic devices, such as the bat detector illustrated above, can be used to estimate population density. In this case, the detector is tuned to the particular frequency of the hunting clicks emitted by the bat species of interest. The number of calls recorded per unit time can be used to estimate numbers within a certain area.

1. Describe what each the following types of sampling equipment is used for in a sampling context:

    (a) Kick sampling technique: _Provides a semi-quantitative sample of substrate-dwelling stream invertebrates_

    (b) Beating tray: _____

    (c) Longworth small mammal trap: _____

    (d) Plankton net: _____

    (e) Sweep net: _____

    (f) Water sampler: _____

    (g) Pooter: _____

    (h) Tullgren funnel: _____

    (i) Pitfall trap: _____

2. Explain why pitfall traps are not recommended for estimates of population density: _____

    _____

3. (a) Explain what influence mesh size might have on the sampling efficiency of a plankton net: _____

    _____

    (b) Explain how this would affect your choice of mesh size when sampling animals in a pond: _____

    _____

    _____

Sampling Using Radio-tracking

Field work involving difficult terrain, aquatic environments, or highly mobile, secretive, or easily disturbed species, has been greatly assisted in recent years by the use of radio-transmitter technology. Radio-tracking is particularly suited to population studies of threatened species (because it is relatively non-invasive) and of pests (because their dispersal and habitat use can be monitored). There are many reasons why radio-tracking is used to follow animal movements. Importantly, radio-tracking can be used to quickly obtain accurate information about an animal's home range. This knowledge is essential to understanding dispersal, distribution, habitat use, and competitive relationships. The information can be used to manage an endangered species effectively, or to plan efficient pest control operations. Satellite transmitters can be used to study the large scale migratory movements of large animals and marine species, which are more difficult to follow using the usual VHF radio-tracking equipment.

Adelie penguins with transmitters

Scanning receiver with 400 channels

Aerial tracking of rooks (a pest bird)

Hand-held portable antenna

A transmitter emits a pulsed signal that can be picked up by a receiver. Transmitters may be short or long range. Each transmitter has an antenna which may be whiplike (above left) or a loop type, which is often incorporated into a collar. Transmitter size and antenna length is set so that there is no interference with the animal's behaviour. Usually the weight of the transmitter does not exceed 2% of the animal's body weight. Receivers pick up the transmitter signal and different channels can be used to display signals from many transmitters. Many are very portable and are used in difficult terrain.

A tracking antenna together with the receiver can be used to home in on an animal. An antenna is directional and so can accurately fix an animal's position. Antennae can be mounted onto light aircraft or off-road vehicles to provide mobile tracking over large areas. For work in inaccessible or difficult terrain, portable, hand-held antennae are used. A folding model is useful in thick scrub and is easily carried when not in use. The more portable the antenna the less powerful it is in terms of picking up the transmitter signal. Very portable antennae are useful only for short range tracking.

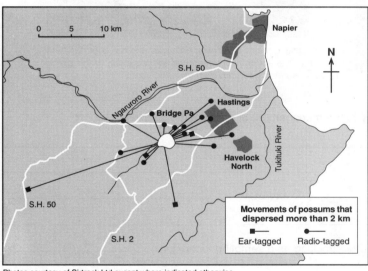

A possum (a major pest in New Zealand) with a transmitter around its neck. The antenna can be seen above the back. Radio-tracking is used on pest species to determine dispersal rates, distribution and habitat use. With this information, pest control can be implemented more effectively. The map to the left shows the dispersal movements of radio-tagged and ear-tagged possums on the east coast of New Zealand.

Photos courtesy of Sirtrack Ltd except where indicated otherwise

Giant weta with transmitter

Long tailed bat with transmitter

Male tuatara wearing transmitter

Goldstripe gecko with transmitter

Radio-tracking equipment is now widely used in many aspects of conservation work. Very small animals, such as weta (left), native bats, and small lizards require very small transmitters, weighing less than 1 g. These are often glued on, or mounted on a body harness. Radio-tracking has been used to study the movements and habitat use of animals as diverse as weta and New Zealand short tailed and long tailed bats (pekapeka). Such knowledge allows DoC to develop better management strategies for these species in the wild.

Tracking devices are used extensively by DoC and university research groups to study NZ's native reptiles. The findings of radio-tracking studies of tuatara indicated that they will use artificial burrows after translocation to offshore islands. The studies also showed and that captive reared individuals have different dispersal patterns to those moved directly from the wild. The niche requirements of the endangered goldstripe gecko were investigated in a similar way, using short range transmitters, located using harmonic radar.

TRANSMITTER SENSING OPTIONS

Transmitters can be set to provide specific information about the activity or physiological state of an animal. Sensing options include:

Activity sensing
The transmitter records changes in posture or behaviour.

Mortality transmitter
Transmitter pulse rate will double (or halve) if it is not moved for a certain time (indicating death).

Heart rate monitors
The transmitter emits a pulse every time the animal's heart beats. It has particular application in studies of animal responses and stress.

Temperature sensing
In these systems, the pulse rate of the transmitter varies according to animal temperature. A calibration curve or a decoding unit on the receiver is used to convert the pulse rate information to a temperature.

Audio transmitters
These pick up animal sounds. Direct observation of the animal is sometimes needed to interpret the sound.

Accurate tracking of movements within a **home range** provides information about habitat use and range requirements. This allows species (such as this New Zealand native rat) to be monitored and managed within their own habitats.

A range of different tracking collars

Tracking migratory movements provides the opportunity to manage and protect threatened species (such as this green turtle) during all phases of their migration.

1. Describe two applications of radio-tracking technology to the management of endangered species:

    (a) _____

    _____

    (b) _____

    _____

2. Discuss the advantages and disadvantages of using radio-tracking as a sampling technique: _____

    _____

    _____

    _____

    _____

3. Explain why radio-tracking is often used for monitoring pest species for which elimination (not conservation) is the goal:

    _____

    _____

    _____

    _____

4. Identify one major constraint of transmitter design and explain why this needs to be considered:

    _____

    _____

Monitoring Change in an Ecosystem

Much of the importance we place on ecosystem change stems ultimately from what we want from that ecosystem. Ecosystems are monitored for changes in their status so that their usefulness can be maintained, whether that use is for agriculture, industry, recreation, or conservation. Never is this so apparent as in the monitoring of aquatic ecosystems. Aquatic environments of all types provide aesthetic pleasure, food, habitat for wildlife, water for industry and irrigation, and potable water. The different uses of aquatic environments demand different standards of **water quality**. For any water body, this is defined in terms of various chemical, physical, and biological characteristics. Together, these factors define the 'health' of the aquatic ecosystem and its suitability for various desirable uses. Water quality is determined by measurement or analysis on-site or in the laboratory. Other methods, involving the use of **indicator species**, can also be used to biologically assess the health of a water body.

Techniques for Monitoring Water Quality

Some aspects of water quality, such as black disk clarity measurements (above), must be made in the field.

The collection of water samples allows many quality measurements to be carried out in the laboratory.

Telemetry stations transmit continuous measurements of the water level of a lake or river to a central control office.

Temperature and dissolved oxygen measurements must be carried out directly in the flowing water.

Water Quality Standards in Aquatic Ecosystems

Water quality variable	Why measured	Standards applied:
Dissolved oxygen	• A requirement for most aquatic life • Indicator of organic pollution • Indicator of photosynthesis (plant growth)	More than 80% saturation **(F, FS, SG)** More than 5 gm^{-3} **(WS)**
Temperature	• Organisms have specific temperature needs • Indicator of mixing processes • Computer modelling examining the uptake and release of nutrients	Less than 25°C **(F)** Less than 3°C change along a stretch of river **(AE, F, FS, SG)**
Conductivity	• Indicator of total salts dissolved in water • Indicator for geothermal input	
pH (acidity)	• Aquatic life protection • Indicator of industrial discharges, mining	Between pH 6 - 9 **(WS)**
Clarity - turbidity - black disk	• Aesthetic appearance • Aquatic life protection • Indicator of catchment condition, land use	Turbidity: 2 NTU Black disk: more than 1.6 m **(AE, CR, A)**
Colour - light absorption	• Aesthetic appearance • Light availability for excessive plant growth • Indicator of presence of organic matter	
Nutrients (Nitrogen and phosphorus)	• Enrichment, excessive plant growth • Limiting factor for plant and algal growth	DIN: less than 0.100 gm^{-3} DRP: less than 0.030 gm^{-3} **(AE, A)** NO_3^-: less than 10 gm^{-3} **(WS)**
Major ions (Mg^{2+}, Ca^{2+}, Na^+, K^+, Cl^-, HCO_3^-, SO_4^{2-})	• Baseline water quality characteristics • Indicator for catchment soil types, geology • Water hardness (magnesium/calcium) • Buffering capacity for pH change (HCO_3^-)	
Organic carbon	• Indicator of organic pollution • Catchment characteristics	BOD: less than 5 gm^{-3} **(AE, CR, A)**
Faecal bacteria	• Indicator of pollution with faecal matter • Disease risk for swimming etc.	ENT: less than 33 cm^{-3} **(CR)** FC: less than 200 cm^{-3}

Fly fishing is a pursuit which demands high water quality.

Spawning salmon require high oxygen levels for egg survival.

Standards refer to specified water uses: **AE** = aquatic ecosystem protection, **A** = aesthetic, **CR** = contact recreation, **SG** = shellfish gathering, **WS** = water supply, **F** = fishery, **FS** = fish spawning, **SW** = stock watering.

Key to abbreviations: NTU = a unit of measurement for turbidity, DIN = dissolved inorganic nitrogen, DRP = dissolved reactive phosphorus, BOD = biochemical oxygen demand, ENT = enterococci, FC = faecal coliform.

1. Explain why dissolved oxygen, temperature, and clarity measurements are made in the field rather than in the laboratory:

_____

_____

Practical Ecology

Code: DA 2

Calculation and Use of Diversity Indices

One of the best ways to determine the health of an ecosystem is to measure the variety (rather than the absolute number) of organisms living in it. Certain species, called **indicator species**, are typical of ecosystems in a particular state (e.g. polluted or pristine). An objective evaluation of an ecosystem's biodiversity can provide valuable insight into its status, particularly if the species assemblages have changed as a result of disturbance.

Diversity can be quantified using a **diversity index (DI)**. Diversity indices attempt to quantify the degree of diversity and identify indicators for environmental stress or degradation. Most indices of diversity are easy to use and they are widely used in ecological work, particularly for monitoring ecosystem change or pollution. One example, which is a derivation of **Simpson's index**, is described below. Other indices produce values ranging between 0 and almost 1. These are more easily interpreted because of the more limited range of values, but no single index offers the "best" measure of diversity: they are chosen on their suitability to different situations.

Simpson's Index for finite populations

This diversity index (DI) is a commonly used inversion of Simpson's index, suitable for finite populations.

$$DI = \frac{N(N-1)}{\Sigma n(n-1)}$$

After Smith and Smith as per IOB.

Where:

- **DI** = Diversity index
- **N** = Total number of individuals (of all species) in the sample
- **n** = Number of individuals of each species in the sample

This index ranges between 1 (low diversity) and infinity. The higher the value, the greater the variety of living organisms. It can be difficult to evaluate objectively without reference to some standard ecosystem measure because the values calculated can, in theory, go to infinity.

Example of species diversity in a stream

The example describes the results from a survey of stream invertebrates. The species have been identified, but this is not necessary in order to calculate diversity as long as the different species can be distinguished. Calculation of the DI using Simpson's index for finite populations is:

Species	No. of individuals
A (Common backswimmer)	12
B (Stonefly larva)	7
C (Silver water beetle)	2
D (Caddis fly larva)	6
E (Water spider)	5
Total number of individuals = 32	

$$DI = \frac{32 \times 31}{(12\times11) + (7\times6) + (2\times1) + (6\times5) + (5\times4)} = \frac{992}{226} = 4.39$$

A stream community with a high macroinvertebrate diversity (above) in contrast to a low diversity stream community (below).

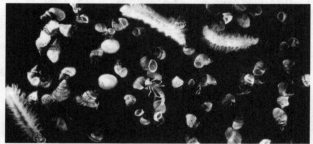

Photos: Stephen Moore

2. Discuss the link between water quality and land use: _____

_____

_____

_____

3. Describe a situation where a species diversity index may provide useful information: _____

_____

_____

4. An area of forest floor was sampled and six invertebrate species were recorded, with counts of 7, 10, 11, 2, 4, and 3 individuals. Using Simpson's index for finite populations, calculate DI for this community:

(a) DI= _____   DI = _____

(b) Comment on the diversity of this community: _____

_____

_____

5. Explain how you could use indicator species to detect pollution in a stream: _____

_____

_____

Using Chi-Squared in Ecology

The **chi-squared test** (χ^2), like the Student's t test, is a test for difference between data sets, but it is used when you are working with frequencies (counts) rather than measurements. It is a simple test to perform but the data must meet the requirements of the test. These are as follows:

■ It can only be used for data that are raw counts (not measurements or derived data such as percentages).

■ It is used to compare an experimental result with an expected theoretical outcome (e.g. an expected Mendelian ratio or a theoretical value indicating "no preference" or "no difference" between groups in some sort of response such as habitat or microclimate preference).

■ It is not a valid test when sample sizes are small (<20).

Like all statistical tests, it aims to test the null hypothesis; the hypothesis of no difference between groups of data. The following exercise is a worked example using chi-squared for testing an ecological study of habitat preference. As with most of these simple statistical tests, chi-squared is easily calculated using a spreadsheet. Guidelines for this are available on Biozone's Teacher Resource CD-ROM.

Using χ^2 in Ecology

In an investigation of the ecological niche of the mangrove, *Avicennia marina var. resinifera*, the density of pneumatophores was measured in regions with different substrate. The mangrove trees were selected from four different areas: mostly sand, some sand, mostly mud, and some mud. Note that the variable, substrate type, is categorical in this case. Quadrats (1 m by 1 m) were placed around a large number of trees in each of these four areas and the numbers of pneumatophores were counted. Chi-squared was used to compare the observed results for pneumatophore density (as follows) to an expected outcome of no difference in density between substrates.

Pneumatophores

Mangrove pneumatophore density in different substrate areas			
Mostly sand	85	Mostly mud	130
Some sand	102	Some mud	123

Using χ^2, the probability of this result being consistent with the expected result could be tested. Worked example as follows:

Step 1: Calculate the expected value (E)

In this case, this is the sum of the observed values divided by the number of categories. $\frac{440}{4} = 110$

Step 2: Calculate O – E

The difference between the observed and expected values is calculated as a measure of the deviation from a predicted result. Since some deviations are negative, they are all squared to give positive values. This step is usually performed as part of a tabulation (right, darker grey column).

Category	O	E	O – E	(O – E)^2	$\frac{(O-E)^2}{E}$
Mostly sand	85	110	–25	625	5.68
Some sand	102	110	–8	64	0.58
Mostly mud	130	110	20	400	3.64
Some mud	123	110	13	169	1.54
	Total = 440				χ^2 $\sum = 11.44$

Step 3: Calculate the value of χ^2

$$\chi^2 = \sum \frac{(O-E)^2}{E}$$

Where: O = the observed result
E = the expected result
$\sum$ = sum of

The calculated χ^2 value is given at the bottom right of the last column in the tabulation.

Step 5a: Using the χ^2 table

On the χ^2 table (part reproduced in Table 1 below) with 3 degrees of freedom, the calculated value for χ^2 of 11.44 corresponds to a probability of between 0.01 and 0.001 (see arrow). *This means that by chance alone a χ^2 value of 11.44 could be expected between 1% and 0.1% of the time.*

Step 4: Calculating degrees of freedom

The probability that any particular χ^2 value could be exceeded by chance depends on the number of degrees of freedom. This is simply *one less than the total number of categories* (this is the number that could vary independently without affecting the last value). *In this case: 4–1 = 3.*

Step 5b: Using the χ^2 table

The probability of between 0.1 and 0.01 is lower than the 0.05 value which is generally regarded as significant. The null hypothesis can be rejected and we have reason to believe that the observed results differ significantly from the expected (at $P = 0.05$).

Table 1: Critical values of χ^2 at different levels of probability. By convention, the critical probability for rejecting the null hypothesis (H_0) is 5%. If the test statistic is less than the tabulated critical value for $P = 0.05$ we cannot reject H_0 and the result is not significant. If the test statistic is greater than the tabulated value for $P = 0.05$ we reject H_0 in favour of the alternative hypothesis.

Degrees of freedom	Level of probability (P)									
	0.98	0.95	0.80	0.50	0.20	0.10	0.05	0.02	0.01	0.001
1	0.001	0.004	0.064	0.455	1.64	2.71	3.84	5.41	6.64	10.83
2	0.040	0.103	0.466	1.386	3.22	4.61	5.99	7.82	9.21	13.82
3	0.185	0.352	1.005	2.366	4.64	6.25	7.82	9.84	11.35	16.27
4	0.429	0.711	1.649	3.357	5.99	7.78	9.49	11.67	13.28	18.47
5	0.752	0.145	2.343	4.351	7.29	9.24	11.07	13.39	15.09	20.52

χ^2 (0.01 column, row 1: 10.83; row 2: 13.82)

← Do not reject H_0 Reject H_0 →

Practical Ecology

Classification

Describing classification systems and using keys

Identifying biological diversity. Distinguishing features of major groups of organisms. Classification systems and keys.

Learning Objectives

☐ 1. Compile your own glossary from the **KEY WORDS** displayed in **bold type** in the learning objectives below.

Classification Systems *(pages 97-108)*

☐ 2. Explain what is meant by **classification**. Recognise **taxonomy** as the study of the theory and practice of classification. Describe the principles and importance of scientific classification.

☐ 3. Describe the **distinguishing features** of each of the kingdoms in the **five kingdom classification system**:
 • **Prokaryotae**: bacteria and cyanobacteria.
 • **Protista**: includes the algae and protozoans.
 • **Fungi**: includes yeasts, moulds, and mushrooms.
 • **Plantae**: includes mosses, liverworts, tracheophytes.
 • **Animalia**: all invertebrate phyla and the chordates.

 Note that the **six kingdom classification system separates out the Prokaryotae** into two separate kingdoms, i.e. **Archaebacteria**: the archaebacteria. and **Eubacteria**: the "true" bacteria.

☐ 4. Recognise at least seven major **taxonomic categories**: **kingdom, phylum, class, order, family, genus**, and **species**. Do not confuse taxonomic categories with **taxa** (sing. **taxon**), which are groups of real organisms: "genus" is a taxonomic category, whereas the genus *Drosophila* is a taxon.

☐ 5. Understand the basis for assigning organisms into different taxonomic categories. Recall what is meant by a **distinguishing feature**. Explain that species are usually classified on the basis of **shared derived characters** rather than primitive (ancestral) characters. *For example, within the subphylum Vertebrata, the presence of a backbone is a derived, therefore a distinguishing, feature. However, within the class Mammalia, the backbone is an ancestral feature and is not distinguishing, whereas mammary glands (a distinguishing feature) are derived.*

☐ 6. Explain how **binomial nomenclature** is used to classify organisms and understand the rules of presentation for these names. Explain the limitations of using **common names** to identify organisms.

Use of Taxonomic Keys *(pages 109-111)*

☐ 7. Demonstrate an understanding of the proper use of **classification keys**. Describe the essential features of any classification key.

☐ 8. Apply and/or design a simple taxonomic key to identify and classify a group of up to eight organisms.

See page 7 for additional details of this text:
■ Smith, R. L. & Smith, T. M., 2001. **Ecology and Field Biology**, reading as required.

See page 7 for details of publishers of periodicals:

STUDENT'S REFERENCE

■ **The Family Line - The Human-Cat Connection** National Geographic, 191(6) June 1997, pp. 77-85. *An examination of the genetic diversity and lineages within the felidae. A good context within which to study classification.*

■ **Biological Keys** The Am. Biology Teacher, 66(3), March 2004, pp. 202-207. *Taxonomic categorization and learning to use biological keys.*

■ **Is it Kingdoms or Domains?** The Am. Biology Teacher, 66(4), April 2004, pp. 268-276. *How many kingdoms should we recognise and how do potentially larger groupings such as domains fit into our picture of kingdom-based classification?*

TEACHER'S REFERENCE

■ **A Universal Phylogenetic Tree** The Am. Biology Teacher, 63(3), March 2001, pp. 164-170. *A comparison of the three domains versus five kingdoms systems, with an examination of the features of the Archaea and the Eubacteria.*

■ **The Species Problem and the Value of Teaching the Complexities of Species** The Am. Biology Teacher, 66(6), Aug. 2004, pp. 413-417. *Problems with the complexities of species definitions and the consequences of these difficulties to species classifications.*

■ **Taxonomy: The Naming Game Revisited** Biol. Sci. Rev., 9(5) May 1997, pp. 31-35. *New tools for taxonomy and how they are used (includes the exemplar of the reclassification of the kingdoms).*

■ **What's in a Name** Scientific American, Nov. 2004, pp. 20-21. *A proposed classification system called phyloclode, based solely on phylogeny.*

■ **Extremophiles** Scientific American, April 1997, pp. 66-71. *The biology and taxonomy of microbial populations of extreme environments.*

■ **Gladiators: A New Order of Insects** Scientific American, Nov. 2002, pp. 42-47. *A new order of insects has recently been described. The article describes their characteristics, distribution, life history, and taxonomy.*

■ **The Loves of the Plants** Scientific American, Feb. 1996, pp. 98-103. *The classification of plants and the development of keys to plant identification.*

See pages 4-5 for details of how to access **Bio Links** from our web site: **www.thebiozone.com** From Bio Links, access sites under the topics:

GENERAL BIOLOGY ONLINE RESOURCES > Online Textbooks and Lecture Notes: • Biology online.org ... *and others* > **General Online Biology Resources**: • Ken's BioWeb resources ...*and others* > **Glossaries**: • Biodiversity glossary

BIODIVERSITY > Taxonomy and Classification: • Birds and DNA • Classification of living things • Introduction to the Eukaryota • Taxonomy: Classifying life • Taxonomy lab • The tree of life homepage • The phylogeny of life... *and others*

MICROBIOLOGY > General Microbiology: • British Mycological Society • Major groups of prokaryotes • The microbial world ... *and others*

PLANT BIOLOGY > Classification and Diversity: • Flowering plant diversity ... *and others*

Presentation MEDIA to support this topic:

ECOLOGY • Biodiversity & Conservation

Features of Taxonomic Groups

In order to distinguish organisms, it is desirable to classify and name them (a science known as **taxonomy**). An effective classification system requires features that are distinctive to a particular group of organisms. The distinguishing features of some major taxonomic groups are provided in the following pages by means of diagrams and brief summaries. Revised classification systems, recognising three domains (rather than five or six kingdoms) are now recognised as better representations of the true diversity of life. However, for the purposes of describing the groups with which we are most familiar, the five kingdom system (used here) is still appropriate. Note that most animals show **bilateral symmetry** (body divisible into two halves that are mirror images). **Radial symmetry** (body divisible into equal halves through various planes) is a characteristic of cnidarians and ctenophores. Definitions of specific terms relating to features of structure or function can be found in any general biology text.

Kingdom: PROKARYOTAE (Bacteria)

- Also known as monerans or prokaryotes.
- Two major bacterial lineages are recognised: the primitive **Archaebacteria** and the more advanced **Eubacteria**.
- All have a prokaryotic cell structure: they lack the nuclei and chromosomes of eukaryotic cells, and have smaller (70S) ribosomes.
- Have a tendency to spread genetic elements across species barriers by sexual conjugation, viral transduction and other processes.
- Can reproduce rapidly by binary fission in the absence of sex.

- Have evolved a wider variety of metabolism types than eukaryotes.
- Bacteria grow and divide or aggregate into filaments or colonies of various shapes.
- They are taxonomically identified by their appearance (form) and through biochemical differences.

Species diversity: 10 000+ Bacteria are rather difficult to classify to species level because of their relatively rampant genetic exchange, and because their reproduction is usually asexual.

Eubacteria

- Also known as 'true bacteria', they probably evolved from the more ancient Archaebacteria.
- Distinguished from Archaebacteria by differences in cell wall composition, nucleotide structure, and ribosome shape.
- Very diverse group comprises most bacteria.
- The **gram stain** provides the basis for distinguishing two broad groups of bacteria. It relies on the presence of peptidoglycan (unique to bacteria) in the cell wall. The stain is easily washed from the thin peptidoglycan layer of gram negative walls but is retained by the thick peptidoglycan layer of gram positive cells, staining them a dark violet colour.

Gram-Positive Bacteria

The walls of gram positive bacteria consist of many layers of peptidoglycan forming a thick, single-layered structure that holds the gram stain.

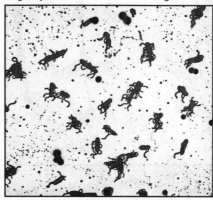

Bacillus alvei: a gram positive, flagellated bacterium. Note how the cells appear dark.

Gram-Negative Bacteria

The cell walls of gram negative bacteria contain only a small proportion of peptidoglycan, so the dark violet stain is not retained by the organisms.

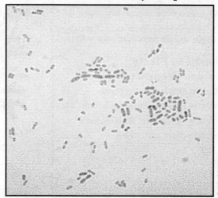

Photos: CDC

Alcaligenes odorans: a gram negative bacterium. Note how the cells appear pale.

Kingdom: FUNGI

- Heterotrophic.
- Rigid cell wall made of chitin.
- Vary from single celled to large multicellular organisms.
- Mostly saprotrophic (ie. feeding on dead or decaying material).
- Terrestrial and immobile.

Examples:
Mushrooms/toadstools, yeasts, truffles, morels, moulds, and lichens.

Species diversity: 80 000 +

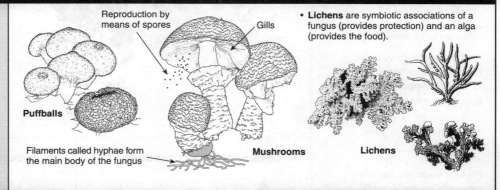

Reproduction by means of spores

Gills

- **Lichens** are symbiotic associations of a fungus (provides protection) and an alga (provides the food).

Puffballs

Filaments called hyphae form the main body of the fungus

Mushrooms

Lichens

Kingdom: PROTISTA

- A diverse group of organisms which do not fit easily into other taxonomic groups.
- Unicellular or simple multicellular.
- Widespread in moist or aquatic environments.

Examples of algae: green, brown, and red algae, dinoflagellates, diatoms.

Examples of protozoa: amoebas, foraminiferans, radiolarians, ciliates.

Species diversity: 55 000 +

Algae 'plant-like' protists

- Autotrophic (photosynthesis)
- Characterised by the type of chlorophyll present

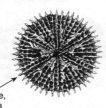

Cell walls of cellulose, sometimes with silica

Diatom

Protozoa 'animal-like' protists

- Heterotrophic nutrition and feed via ingestion
- Most are microscopic (5 µm - 250 µm)

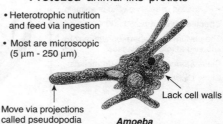

Move via projections called pseudopodia

Lack cell walls

Amoeba

Classification

Kingdom: PLANTAE

- Multicellular organisms (the majority are photosynthetic and contain chlorophyll).
- Cell walls made of cellulose; Food is stored as starch.
- Subdivided into two major divisions based on tissue structure: **Bryophytes** (non-vascular) and **Tracheophytes** (vascular) plants.

Non-Vascular Plants:

- Non vascular, lacking transport tissues (no xylem or phloem).
- They are small and restricted to moist, terrestrial environments.
- Do not possess 'true' roots, stems or leaves.

Phylum Bryophyta: Mosses, liverworts, and hornworts.

Species diversity: 18 600 +

Phylum: Bryophyta

Sexual reproductive structures

Flattened thallus (leaf like structure)

Sporophyte: reproduce by spores

Rhizoids anchor the plant into the ground

Liverworts

Mosses

Vascular Plants:

- Vascular: possess transport tissues.
- Possess true roots, stems, and leaves, as well as stomata.
- Reproduce via spores, not seeds.
- Clearly defined *alternation of sporophyte and gametophyte generations*.

Seedless Plants:

Spore producing plants, includes:

Phylum Filicinophyta: Ferns
Phylum Sphenophyta: Horsetails
Phylum Lycophyta: Club mosses
Species diversity: 13 000 +

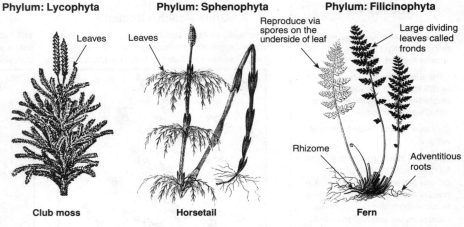

Phylum: Lycophyta
Leaves

Club moss

Phylum: Sphenophyta
Leaves

Horsetail

Phylum: Filicinophyta
Reproduce via spores on the underside of leaf

Large dividing leaves called fronds

Rhizome

Adventitious roots

Fern

Seed Plants:

Also called Spermatophyta. Produce seeds housing an embryo. Includes:

Gymnosperms

- Lack enclosed chambers in which seeds develop.
- Produce seeds in cones which are exposed to the environment.

Phylum Cycadophyta: Cycads
Phylum Ginkgophyta: Ginkgoes
Phylum Coniferophyta: Conifers
Species diversity: 730 +

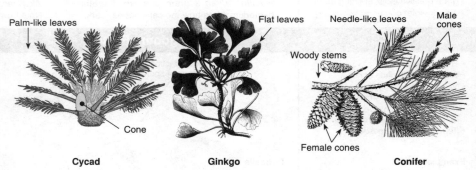

Phylum: Cycadophyta
Palm-like leaves

Cone

Cycad

Phylum: Ginkophyta
Flat leaves

Ginkgo

Phylum: Coniferophyta
Needle-like leaves

Male cones

Woody stems

Female cones

Conifer

Angiosperms

Phylum: Angiospermophyta

- Seeds in specialised reproductive structures called flowers.
- Female reproductive ovary develops into a fruit.
- Pollination usually via wind or animals.

Species diversity: 260 000 +

The phylum Angiospermophyta may be subdivided into two classes:

Class *Monocotyledoneae* (Monocots)
Class *Dicotyledoneae* (Dicots)

Angiosperms: **Monocotyledons**

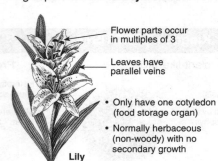

Flower parts occur in multiples of 3

Leaves have parallel veins

- Only have one cotyledon (food storage organ)
- Normally herbaceous (non-woody) with no secondary growth

Lily

Examples: cereals, lilies, daffodils, palms, grasses.

Angiosperms: **Dicotyledons**

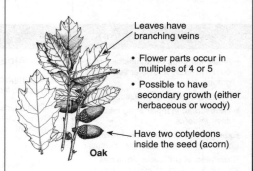

Leaves have branching veins

- Flower parts occur in multiples of 4 or 5
- Possible to have secondary growth (either herbaceous or woody)

Have two cotyledons inside the seed (acorn)

Oak

Examples: many annual plants, trees and shrubs.

Kingdom: ANIMALIA

- Over 800 000 species described in 33 existing phyla.
- Multicellular, heterotrophic organisms.
- Animal cells lack cell walls.

- Further subdivided into various major phyla on the basis of body symmetry, type of body cavity, and external and internal structures.

Phylum: Rotifera

- A diverse group of small organisms with sessile, colonial, and planktonic forms.
- Most freshwater, a few marine.
- Typically reproduce via cyclic parthenogenesis.
- Characterised by a wheel of cilia on the head used for feeding and locomotion, a large muscular pharynx (mastax) with jaw like trophi, and a foot with sticky toes.

Species diversity: 1500 +

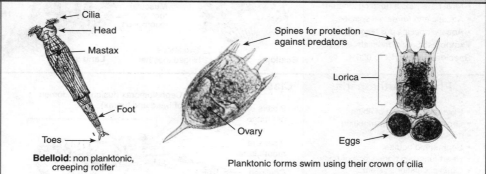

Cilia
Head
Mastax
Foot
Toes

Bdelloid: non planktonic, creeping rotifer

Spines for protection against predators

Lorica

Ovary

Eggs

Planktonic forms swim using their crown of cilia

Phylum: Porifera

- Lack organs.
- All are aquatic (mostly marine).
- Asexual reproduction by budding.
- Lack a nervous system.

Examples: sponges.

Species diversity: 8000 +

Body wall perforated by pores through which water enters

Water leaves by a larger opening - the osculum

Sponge

- Capable of regeneration (the replacement of lost parts)
- Possess spicules (needle-like internal structures) for support and protection

Tube sponge

Sessile (attach to ocean floor)

Phylum: Cnidaria

- Two basic body forms:

  Medusa: umbrella shaped and free swimming by pulsating bell.

  Polyp: cylindrical, some are sedentary, others can glide, or somersault or use tentacles as legs.

- Some species have a life cycle that alternates between a polyp stage and a medusa stage.
- All are aquatic (most are marine).

Examples: Jellyfish, sea anemones, hydras, and corals.

Species diversity: 11 000 +

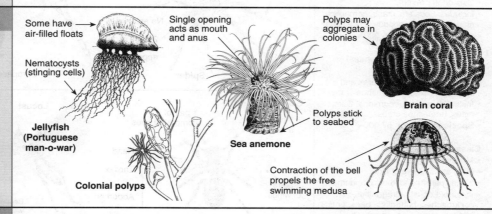

Some have air-filled floats

Nematocysts (stinging cells)

Jellyfish (Portuguese man-o-war)

Colonial polyps

Single opening acts as mouth and anus

Sea anemone

Polyps may aggregate in colonies

Brain coral

Polyps stick to seabed

Contraction of the bell propels the free swimming medusa

Phylum: Platyhelminthes

- Unsegmented body.
- Flattened body shape.
- Mouth, but no anus.
- Many are parasitic.

Examples: Tapeworms, planarians, flukes.

Species diversity: 20 000 +

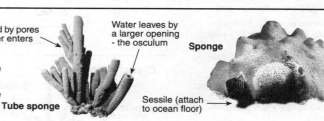

Hooks

Detail of head (scolex)

Liver fluke

Tapeworm

Planarian

Phylum: Nematoda

- Tiny, unsegmented roundworms.
- Many are plant/animal parasites

Examples: Hookworms, stomach worms, lung worms, filarial worms

Species diversity: 80 000 - 1 million

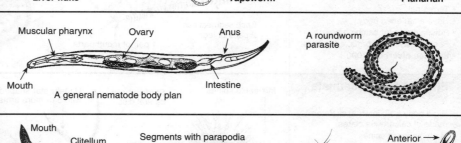

Muscular pharynx
Ovary
Anus

A roundworm parasite

Mouth
Intestine

A general nematode body plan

Phylum: Annelida

- Cylindrical, segmented body with chaetae (bristles).
- Move using hydrostatic skeleton and/or parapodia (appendages).

Examples: Earthworms, leeches, polychaetes (including tubeworms).

Species diversity: 15 000 +

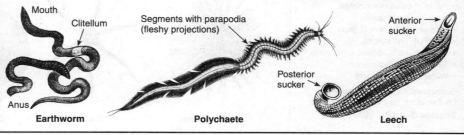

Mouth
Clitellum

Segments with parapodia (fleshy projections)

Anterior sucker

Posterior sucker

Anus
Earthworm

Polychaete

Leech

Classification

Kingdom: ANIMALIA *(continued)*

Phylum: Mollusca

- Soft bodied and unsegmented.
- Body comprises head, muscular foot, and visceral mass (organs).
- Most have radula (rasping tongue).
- Aquatic and terrestrial species.
- Aquatic species possess gills.

Examples: Snails, mussels, squid.
Species diversity: 110 000 +

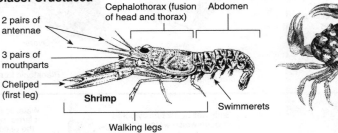

Class: Bivalvia

Radula lost in bivalves
Mantle secretes shell
Muscular foot for locomotion
Two shells hinged together
Scallop

Class: Gastropoda

Mantle secretes shell
Tentacles with eyes
Head
Muscular foot for locomotion
Land snail

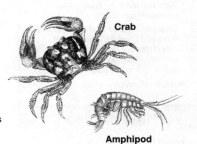

Class: Cephalopoda

Well developed eyes
Squid
Foot divided into tentacles

Phylum: Arthropoda

- Exoskeleton made of chitin.
- Grow in stages after moulting.
- Jointed appendages.
- Segmented bodies.
- Heart found on dorsal side of body.
- Open circulation system.
- Most have compound eyes.

Species diversity: 1 million +
Make up 75% of all living animals.

Arthropods are subdivided into the following classes:

Class: Crustacea (crustaceans)
- Mainly marine.
- Exoskeleton impregnated with mineral salts.
- Gills often present.
- Includes: Lobsters, crabs, barnacles, prawns, shrimps, isopods, amphipods
- **Species diversity:** 35 000 +

Class: Arachnida (chelicerates)
- Almost all are terrestrial.
- 2 body parts: cephalothorax and abdomen (except horseshoe crabs).
- Includes: spiders, scorpions, ticks, mites, horseshoe crabs.
- **Species diversity:** 57 000 +

Class: Insecta (insects)
- Mostly terrestrial.
- Most are capable of flight.
- 3 body parts: head, thorax, abdomen.
- Include: Locusts, dragonflies, cockroaches, butterflies, bees, ants, beetles, bugs, flies, and more
- **Species diversity:** 800 000 +

Myriapods (=many legs)
Class Diplopoda (millipedes)
- Terrestrial.
- Have a rounded body.
- Eat dead or living plants.
- **Species diversity:** 2000 +

Class Chilopoda (centipedes)
- Terrestrial.
- Have a flattened body.
- Poison claws for catching prey.
- Feed on insects, worms, and snails.
- **Species diversity:** 7000 +

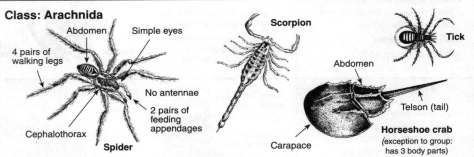

Class: Crustacea

2 pairs of antennae
Cephalothorax (fusion of head and thorax)
Abdomen
Crab
3 pairs of mouthparts
Cheliped (first leg)
Shrimp
Swimmerets
Walking legs
Amphipod

Class: Arachnida

Abdomen
Simple eyes
Scorpion
Tick
4 pairs of walking legs
No antennae
2 pairs of feeding appendages
Abdomen
Cephalothorax
Spider
Carapace
Telson (tail)
Horseshoe crab
(exception to group: has 3 body parts)

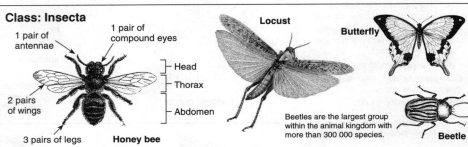

Class: Insecta

1 pair of antennae
1 pair of compound eyes
Locust
Butterfly
Head
Thorax
Abdomen
2 pairs of wings
3 pairs of legs
Honey bee
Beetles are the largest group within the animal kingdom with more than 300 000 species.
Beetle

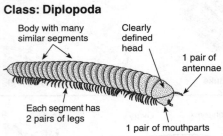
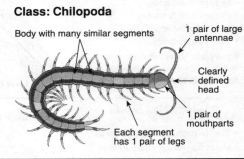

Class: Diplopoda

Body with many similar segments
Clearly defined head
1 pair of antennae
Each segment has 2 pairs of legs
1 pair of mouthparts

Class: Chilopoda

Body with many similar segments
1 pair of large antennae
Clearly defined head
1 pair of mouthparts
Each segment has 1 pair of legs

Phylum: Echinodermata

- Rigid body wall, internal skeleton made of calcareous plates.
- Many possess spines.
- Ventral mouth, dorsal anus.
- External fertilisation.
- Unsegmented, marine organisms.
- Tube feet for locomotion.
- Water vascular system.

Examples: Starfish, brittlestars, feather stars, sea urchins, sea lilies.
Species diversity: 6000 +

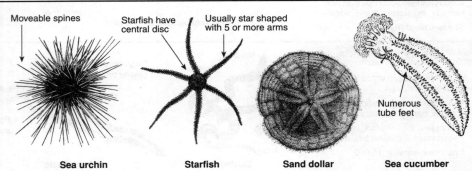

Moveable spines
Starfish have central disc
Usually star shaped with 5 or more arms
Numerous tube feet
Sea urchin
Starfish
Sand dollar
Sea cucumber

Kingdom: ANIMALIA (continued)

Phylum: Chordata

- Dorsal notochord (flexible, supporting rod) present at some stage in the life history.
- Post-anal tail present at some stage in their development.
- Dorsal, tubular nerve cord.
- Pharyngeal slits present.
- Circulation system closed in most.
- Heart positioned on ventral side.

Species diversity: 48 000 +

- A very diverse group with several sub-phyla:
  - Urochordata (sea squirts, salps)
  - Cephalochordata (lancelet)
  - Craniata (vertebrates)

Sub-Phylum Craniata (vertebrates)
- Internal skeleton of cartilage or bone.
- Well developed nervous system.
- Vertebral column replaces notochord.
- Two pairs of appendages (fins or limbs) attached to girdles.

Further subdivided into:

Class: Chondrichthyes (cartilaginous fish)
- Skeleton of cartilage (not bone).
- No swim bladder.
- All aquatic (mostly marine).
- Include: Sharks, rays, and skates.

Species diversity: 850 +

Class: Osteichthyes (bony fish)
- Swim bladder present.
- All aquatic (marine and fresh water).

Species diversity: 21 000 +

Class: Amphibia (amphibians)
- Lungs in adult, juveniles may have gills (retained in some adults).
- Gas exchange also through skin.
- Aquatic and terrestrial (limited to damp environments).
- Include: Frogs, toads, salamanders, and newts.

Species diversity: 3900 +

Class Reptilia (reptiles)
- Ectotherms with no larval stages.
- Teeth are all the same type.
- Eggs with soft leathery shell.
- Mostly terrestrial.
- Include: Snakes, lizards, crocodiles, turtles, and tortoises.

Species diversity: 7000 +

Class: Aves (birds)
- Terrestrial endotherms.
- Eggs with hard, calcareous shell.
- Strong, light skeleton.
- High metabolic rate.
- Gas exchange assisted by air sacs.

Species diversity: 8600 +

Class: Mammalia (mammals)
- Endotherms with hair or fur.
- Mammary glands produce milk.
- Glandular skin with hair or fur.
- External ear present.
- Teeth are of different types.
- Diaphragm between thorax/abdomen.

Species diversity: 4500 +
Subdivided into three subclasses: *Monotremes, marsupials, placentals.*

Class: Chondrichthyes (cartilaginous fish)

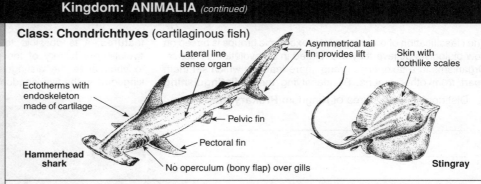

Ectotherms with endoskeleton made of cartilage
Lateral line sense organ
Asymmetrical tail fin provides lift
Skin with toothlike scales
Pelvic fin
Pectoral fin
No operculum (bony flap) over gills
Hammerhead shark
Stingray

Class: Osteichthyes (bony fish)

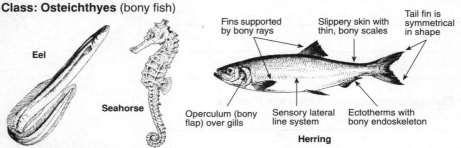

Eel
Seahorse
Fins supported by bony rays
Slippery skin with thin, bony scales
Tail fin is symmetrical in shape
Operculum (bony flap) over gills
Sensory lateral line system
Ectotherms with bony endoskeleton
Herring

Class: Amphibia

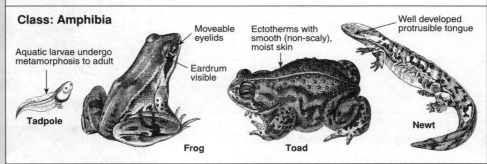

Aquatic larvae undergo metamorphosis to adult
Moveable eyelids
Eardrum visible
Ectotherms with smooth (non-scaly), moist skin
Well developed protrusible tongue
Tadpole
Frog
Toad
Newt

Class: Reptilia

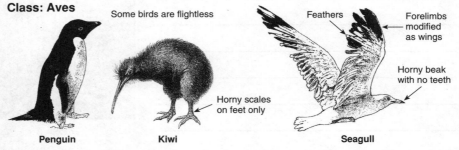

Protective shell of horny plates
Dry, watertight skin covered with overlapping scales
Most with well developed eyes
Limbs absent
Crocodile
Sea turtle
Rattlesnake

Class: Aves

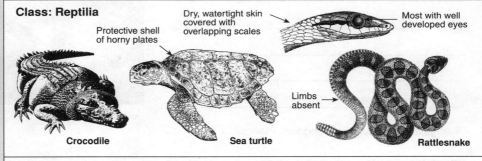

Some birds are flightless
Feathers
Forelimbs modified as wings
Horny beak with no teeth
Horny scales on feet only
Penguin
Kiwi
Seagull

Class: Mammalia

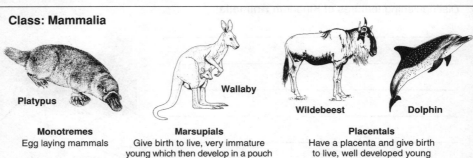

Platypus
Wallaby
Wildebeest
Dolphin
Monotremes
Egg laying mammals
Marsupials
Give birth to live, very immature young which then develop in a pouch
Placentals
Have a placenta and give birth to live, well developed young

Classification

Features of the Five Kingdoms

The classification of organisms into taxonomic groups is based on how biologists believe they are related in an evolutionary sense. Organisms in a taxonomic group share features which set them apart from other groups. By identifying these **distinguishing** features, it is possible to develop an understanding of the evolutionary history of the group. The focus of this activity is to summarise the distinguishing features of each of the five kingdoms in the five kingdom classification system.

1. Distinguishing features of Kingdom **Prokaryotae**:

_____

_____

_____

_____

_____

_____

2. Distinguishing features of Kingdom **Protista**:

_____

_____

_____

_____

_____

_____

3. Distinguishing features of Kingdom **Fungi**:

_____

_____

_____

_____

_____

4. Distinguishing features of Kingdom **Plantae**:

_____

_____

_____

_____

_____

5. Distinguishing features of Kingdom **Animalia**:

_____

_____

_____

_____

_____

Spirillum bacteria

Staphylococcus

Foraminiferan

Spirogyra algae

Mushrooms

Yeast cells in solution

Moss

Pea plants

Cicada moulting

Gibbon

Features of Microbial Groups

A microorganism (or microbe) is literally a microscopic organism. The term is usually reserved for the organisms studied in microbiology: bacteria, fungi, microscopic protistans, and viruses. Most of these taxa also have macroscopic representatives. This is especially the case within the fungi. The distinction between macrofungi and microfungi is an artificial but convenient one.

Unlike microfungi, which are made conspicuous by the diseases or decay they cause, macrofungi are the ones most likely to be observed with the naked eye. Examples of microfungi, which include yeasts and pathogenic species, are illustrated in this activity. Macrofungi, which include mushrooms, toadstools, and lichens, are illustrated in *Features of Macrofungi and Plants*.

1. Distinguishing features of Kingdom **Prokaryotae**:

_____

_____

_____

_____

_____

_____

_____

_____

_____

2. Distinguishing features of Kingdom **Protista**:

_____

_____

_____

_____

_____

_____

_____

_____

3. Distinguishing features of Kingdom **Fungi** (microfungi):

_____

_____

_____

_____

_____

_____

_____

_____

_____

Spirillum bacteria

Staphylococcus

Anabaena cyanobacterium

Foraminiferan

Spirogyra algae

Diatoms: *Pleurosigma*

Curvularia sp. conidiophore

Yeast cells in solution

Microsporum distortum (a pathogenic fungus)

Classification

Code: ERA 1

Features of Macrofungi and Plants

Although plants and fungi are some of the most familiar organisms in our environment, their classification has not always been straightforward. We know now that the plant kingdom is monophyletic, meaning that it is derived from a common ancestor. The variety we see in plant taxa today is a result of their enormous diversification from the first plants. Although the fungi were once grouped together with the plants, they are unique organisms that differ from other eukaryotes in their mode of nutrition, structural organisation, growth, and reproduction. The focus of this activity is to summarise the features of the fungal kingdom, the major divisions of the plant kingdom, and the two classes of flowering plants (angiosperms).

Lichen · Bracket fungus · Liverwort · Moss · Fern frond · Ground fern · Pine tree cone · Cycad · Coconut palms · Wheat plants · Deciduous tree · Flowering plant

1. **Macrofungi** features: _____

2. **Moss** and **liverwort** features: _____

3. **Fern** features: _____

4. **Gymnosperm** features: _____

5. **Monocot angiosperm** features: _____

6. **Dicot angiosperm** features: _____

Features of Animal Taxa

The animal kingdom is classified into about 35 major **phyla**. Representatives of the more familiar taxa are illustrated below: **cnidarians** (includes jellyfish, sea anemones, and corals), **annelids** (segmented worms), **arthropods** (insects, crustaceans, spiders, scorpions, centipedes and millipedes), **molluscs** (snails, bivalve shellfish, squid and octopus), **echinoderms** (starfish and sea urchins), **vertebrates** from the phylum **chordates** (fish, amphibians, reptiles, birds, and mammals). The **arthropods** and the **vertebrates** have been represented in more detail, giving the **classes** for each of these **phyla**. This activity asks you to describe the **distinguishing features** of each of the taxa represented below.

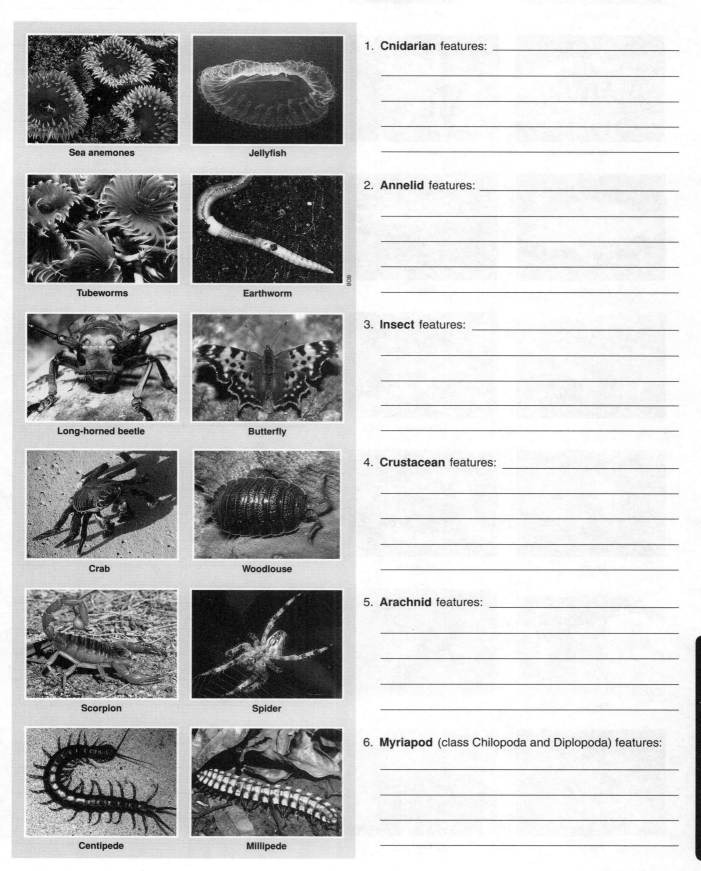

Sea anemones

Jellyfish

Tubeworms

Earthworm

Long-horned beetle

Butterfly

Crab

Woodlouse

Scorpion

Spider

Centipede

Millipede

1. **Cnidarian** features: _____

_____

_____

_____

2. **Annelid** features: _____

_____

_____

_____

3. **Insect** features: _____

_____

_____

_____

4. **Crustacean** features: _____

_____

_____

_____

5. **Arachnid** features: _____

_____

_____

_____

6. **Myriapod** (class Chilopoda and Diplopoda) features:

_____

_____

_____

Classification

Code: R 1

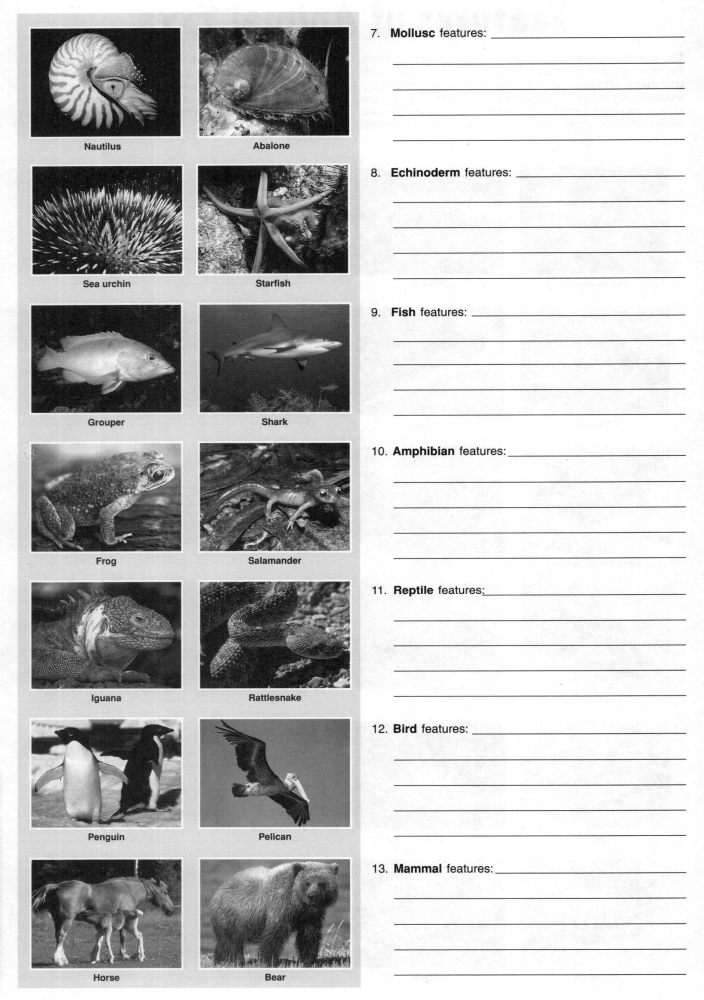

Nautilus

Abalone

Sea urchin

Starfish

Grouper

Shark

Frog

Salamander

Iguana

Rattlesnake

Penguin

Pelican

Horse

Bear

7. **Mollusc** features: _____
_____
_____
_____
_____

8. **Echinoderm** features: _____
_____
_____
_____
_____

9. **Fish** features: _____
_____
_____
_____
_____

10. **Amphibian** features: _____
_____
_____
_____
_____

11. **Reptile** features: _____
_____
_____
_____
_____

12. **Bird** features: _____
_____
_____
_____
_____

13. **Mammal** features: _____
_____
_____
_____
_____

Classification System

The classification of organisms is designed to reflect how they are related to each other. The fundamental unit of classification of living things is the **species**. Its members are so alike genetically that they can interbreed. This genetic similarity also means that they are almost identical in their physical and other characteristics. Species are classified further into larger, more comprehensive categories (higher taxa). It must be emphasised that all such higher classifications are human inventions to suit a particular purpose.

1. The table below shows part of the classification for humans using the seven major levels of classification. For this question, use the example of the classification of the Ethiopian hedgehog, on the following page, as a guide.

    (a) Complete the list of the classification levels on the left hand side of the table below:

Classification level	Human classification
1. _____	_____
2. _____	_____
3. _____	_____
4. _____	_____
5. Family	Hominidae
6. _____	_____
7. _____	_____

    (b) The name of the Family that humans belong to has already been entered into the space provided. Complete the classification for humans (*Homo sapiens*) on the table above.

2. Describe the two-part scientific naming system (called the **binomial system**) that is used to name organisms:

    _____

    _____

3. Give two reasons explaining why the classification of organisms is important:

    (a) _____

    _____

    (b) _____

    _____

4. Traditionally, the classification of organisms has been based largely on similarities in physical appearance. More recently, new methods involving biochemical comparisons have been used to provide new insights into how species are related. Describe an example of a biochemical method for comparing how species are related:

    _____

    _____

    _____

5. As an example of physical features being used to classify organisms, mammals have been divided into three major subclasses: monotremes, marsupials, and placentals. Describe the main physical feature distinguishing each of these taxa:

    (a) Monotreme: _____

    _____

    (b) Marsupial: _____

    _____

    (c) Placental: _____

    _____

Classification

Code: RA 2

Classification of the Ethiopian Hedgehog

Below is the classification for the **Ethiopian hedgehog**. Only one of each group is subdivided in this chart showing the levels that can be used in classifying an organism. Not all possible subdivisions have been shown here. For example, it is possible to indicate such categories as **super-class** and **sub-family**. The only natural category is the **species**, often separated into geographical **races**, or **sub-species**, which generally differ in appearance.

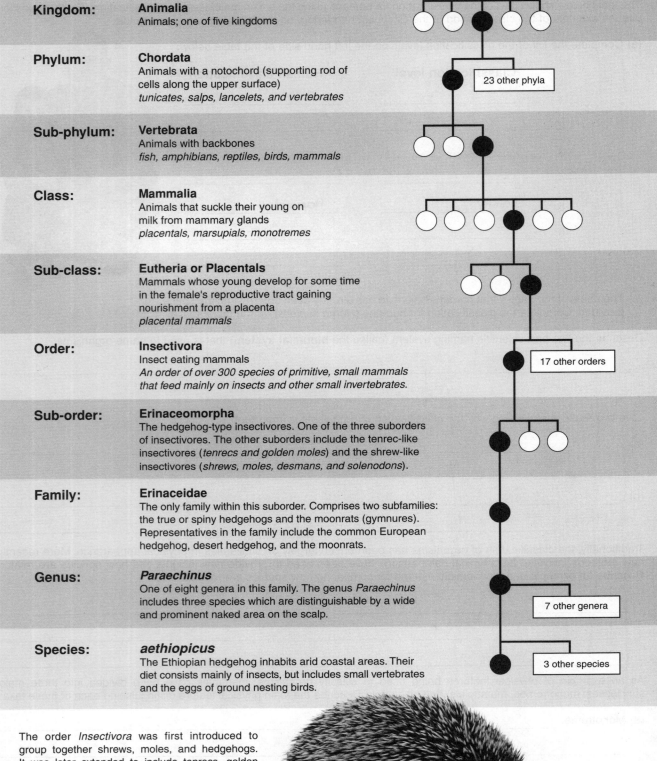

Kingdom: Animalia
Animals; one of five kingdoms

Phylum: Chordata
Animals with a notochord (supporting rod of cells along the upper surface)
tunicates, salps, lancelets, and vertebrates
23 other phyla

Sub-phylum: Vertebrata
Animals with backbones
fish, amphibians, reptiles, birds, mammals

Class: Mammalia
Animals that suckle their young on milk from mammary glands
placentals, marsupials, monotremes

Sub-class: Eutheria or Placentals
Mammals whose young develop for some time in the female's reproductive tract gaining nourishment from a placenta
placental mammals

Order: Insectivora
Insect eating mammals
An order of over 300 species of primitive, small mammals that feed mainly on insects and other small invertebrates.
17 other orders

Sub-order: Erinaceomorpha
The hedgehog-type insectivores. One of the three suborders of insectivores. The other suborders include the tenrec-like insectivores (*tenrecs and golden moles*) and the shrew-like insectivores (*shrews, moles, desmans, and solenodons*).

Family: Erinaceidae
The only family within this suborder. Comprises two subfamilies: the true or spiny hedgehogs and the moonrats (gymnures). Representatives in the family include the common European hedgehog, desert hedgehog, and the moonrats.

Genus: *Paraechinus*
One of eight genera in this family. The genus *Paraechinus* includes three species which are distinguishable by a wide and prominent naked area on the scalp.
7 other genera

Species: *aethiopicus*
The Ethiopian hedgehog inhabits arid coastal areas. Their diet consists mainly of insects, but includes small vertebrates and the eggs of ground nesting birds.
3 other species

The order *Insectivora* was first introduced to group together shrews, moles, and hedgehogs. It was later extended to include tenrecs, golden moles, desmans, tree shrews, and elephant shrews and the taxonomy of the group became very confused. Recent reclassification of the elephant shrews and tree shrews into their own separate orders has made the Insectivora a more cohesive group taxonomically.

Ethiopian hedgehog
Paraechinus aethiopicus

Classification Keys

Classification systems provide biologists with a way in which to identify species. They also indicate how closely related, in an evolutionary sense, each species is to others. An organism's classification should include a clear, unambiguous **description**, an accurate **diagram**, and its unique name, denoted by the **genus** and **species**. Classification keys are used to identify an organism and assign it to the correct species (assuming that the organism has already been formally classified and is included in the key). Typically, keys are **dichotomous** and involve a series of linked steps. At each step, a choice is made between two features; each alternative leads to another question until an identification is made. If the organism cannot be identified, it may be a new species or the key may need revision. Two examples of **dichotomous keys** are provided here. The first (below) describes features for identifying the larvae of various genera within the order Trichoptera (caddisflies). From this key you should be able to assign a generic name to each of the caddisfly larvae pictured. The key on the following page identifies aquatic insect orders.

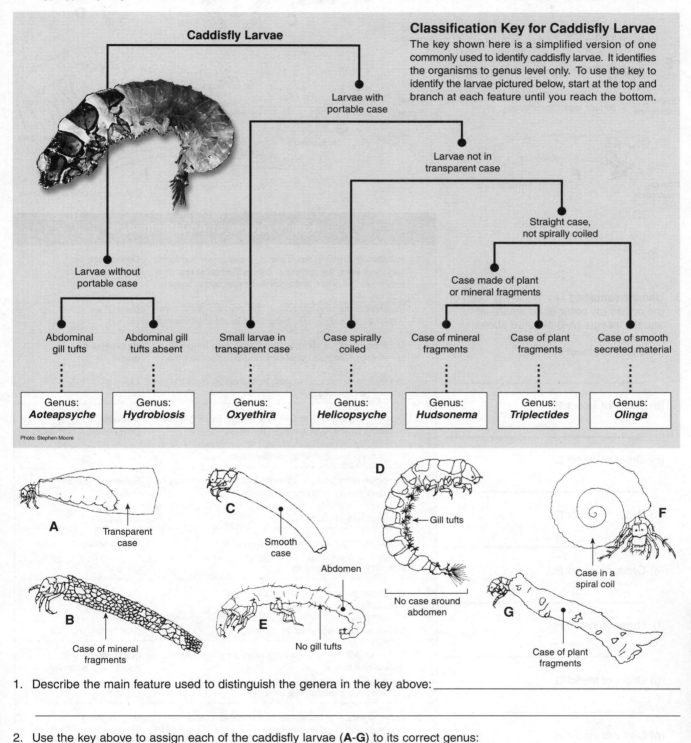

Classification Key for Caddisfly Larvae

The key shown here is a simplified version of one commonly used to identify caddisfly larvae. It identifies the organisms to genus level only. To use the key to identify the larvae pictured below, start at the top and branch at each feature until you reach the bottom.

1. Describe the main feature used to distinguish the genera in the key above: _____

_____

2. Use the key above to assign each of the caddisfly larvae (**A-G**) to its correct genus:

A: _____ D: _____ G: _____

B: _____ E: _____

C: _____ F: _____

Code: A 2

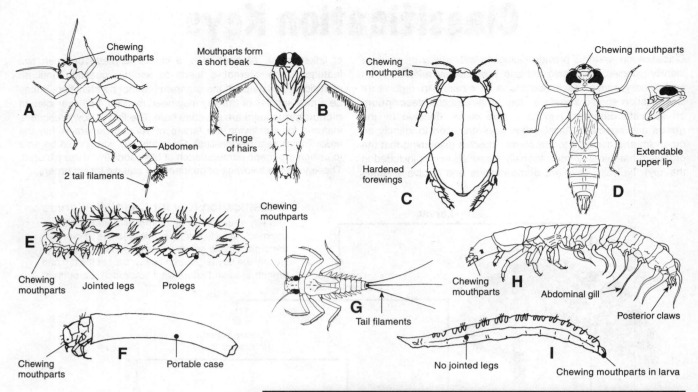

A — Chewing mouthparts; Abdomen; 2 tail filaments

B — Mouthparts form a short beak; Fringe of hairs

C — Chewing mouthparts; Hardened forewings

D — Chewing mouthparts; Extendable upper lip

E — Chewing mouthparts; Jointed legs; Prolegs

F — Chewing mouthparts; Portable case

G — Chewing mouthparts; Tail filaments

H — Chewing mouthparts; Abdominal gill; Posterior claws

I — No jointed legs; Chewing mouthparts in larva

3. Use the simplified key to identify each of the orders (by order or common name) of aquatic insects (**A-I**) pictured above:

(a) Order of insect A:

_____

(b) Order of insect B:

_____

(c) Order of insect C:

_____

(d) Order of insect D:

_____

(e) Order of insect E:

_____

(f) Order of insect F:

_____

(g) Order of insect G:

_____

(h) Order of insect H:

_____

(i) Order of insect I:

_____

Key to Orders of Aquatic Insects

1	Insects with chewing mouthparts; forewings are hardened and meet along the midline of the body when at rest (they may cover the entire abdomen or be reduced in length).	**Coleoptera** (beetles)
	Mouthparts piercing or sucking and form a pointed cone	*Go to 2*
	With chewing mouthparts, but without hardened forewings	*Go to 3*
2	Mouthparts form a short, pointed beak; legs fringed for swimming or long and spaced for suspension on water.	**Hemiptera** (bugs)
	Mouthparts do not form a beak; legs (if present) not fringed or long, or spaced apart.	*Go to 3*
3	Prominent upper lip (labium) extendable, forming a food capturing structure longer than the head.	**Odonata** (dragonflies & damselflies)
	Without a prominent, extendable labium	*Go to 4*
4	Abdomen terminating in three tail filaments which may be long and thin, or with fringes of hairs.	**Ephemeroptera** (mayflies)
	Without three tail filaments	*Go to 5*
5	Abdomen terminating in two tail filaments	**Plecoptera** (stoneflies)
	Without long tail filaments	*Go to 6*
6	With three pairs of jointed legs on thorax	*Go to 7*
	Without jointed, thoracic legs (although non-segmented prolegs or false legs may be present).	**Diptera** (true flies)
7	Abdomen with pairs of non-segmented prolegs bearing rows of fine hooks.	**Lepidoptera** (moths and butterflies)
	Without pairs of abdominal prolegs	*Go to 8*
8	With eight pairs of finger-like abdominal gills; abdomen with two pairs of posterior claws.	**Megaloptera** (dobsonflies)
	Either, without paired, abdominal gills, or, if such gills are present, without posterior claws.	*Go to 9*
9	Abdomen with a pair of posterior prolegs bearing claws with subsidiary hooks; sometimes a portable case.	**Trichoptera** (caddisflies)

© Biozone International 2006
Photocopying Prohibited

Keying Out Plant Species

Dichotomous keys are a useful tool in biology and can enable identification to the species level provided the characteristics chosen are appropriate for separating species. Keys are extensively used by botanists as they are quick and easy to use in the field, although they sometimes rely on the presence of particular plant parts such as fruits or flowers. Some also require some specialist knowledge of plant biology. The following simple activity requires you to identify five species of the genus *Acer* from illustrations of the leaves. It provides valuable practice in using characteristic features to identify plants to species level.

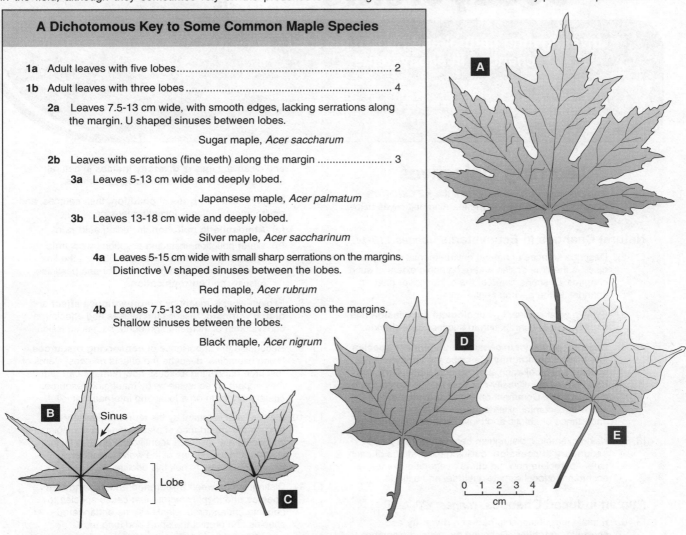

A Dichotomous Key to Some Common Maple Species

1a Adult leaves with five lobes...2

1b Adult leaves with three lobes 4

    2a Leaves 7.5-13 cm wide, with smooth edges, lacking serrations along the margin. U shaped sinuses between lobes.

        Sugar maple, *Acer saccharum*

    2b Leaves with serrations (fine teeth) along the margin 3

      3a Leaves 5-13 cm wide and deeply lobed.

        Japansese maple, *Acer palmatum*

      3b Leaves 13-18 cm wide and deeply lobed.

        Silver maple, *Acer saccharinum*

    4a Leaves 5-15 cm wide with small sharp serrations on the margins. Distinctive V shaped sinuses between the lobes.

        Red maple, *Acer rubrum*

    4b Leaves 7.5-13 cm wide without serrations on the margins. Shallow sinuses between the lobes.

        Black maple, *Acer nigrum*

1. Use the dichotomous key to the common species of *Acer* to identify the species illustrated by the leaves (drawn to scale). Begin at the top of the key and make a choice as to which of the illustrations best fits the description:

(a) Species A: _____

(b) Species B: _____

(c) Species C: _____

(d) Species D: _____

(e) Species E: _____

2. Identify a feature that could be used to identify maple species when leaves are absent: _____

3. Suggest why it is usually necessary to consider a number of different features in order to classify plants to species level:

_____

_____

4. When identifying a plant, suggest what you should be sure of before using a key to classify it to species level:

_____

_____

Classification

Code: A 2

Changes in Ecosystems

Investigating natural and human Induced changes in ecosystems

Environmental changes in natural and modified ecosystems: ecosystem stability, ecological succession, environmental issues.

Learning Objectives

☐ 1. Compile your own glossary from the **KEY WORDS** displayed in **bold type** in the learning objectives below.

Natural Changes in Ecosystems *(pages 113-119)*

☐ 2. Describe sources of natural environmental change and discuss the time scales and geographic extent of such changes. Describe some of the evidence for past changes in the environment.

☐ 3. Explain what is meant by **ecological succession** and identify factors contributing to successional change.

☐ 4. Describe **primary succession** from **pioneer species** to a **climax community**. Describe the characteristics of species typical of each successional stage. Describe how community diversity changes during the course of a succession. Comment on the stability of the pioneer and climax communities and relate this to the relative importance of abiotic and biotic factors at each stage.

☐ 5. Using examples, distinguish between **primary** and **secondary succession**, outlining the features of each type. Appreciate how the **climax** vegetation varies according to local climate, latitude, and altitude.

Human Induced Changes *(pages 120-123)*

☐ 6. Explain the relationship between **diversity** and ecosystem **stability**. Describe the uses, advantages, and disadvantages of **diversity indices** in natural ecosystems and in those modified by humans.

☐ 7. Describe different types of **pollution**, their sources, and environmental effects. Include reference to:

(a) **Atmospheric pollution**, including **acid rain**.

(b) Water pollution, including pollution by **organic effluent** and **fertiliser run-off**. Discuss the link between soil degradation, fertiliser and pesticide misuse, and **eutrophication**.

☐ 8. Explain what is meant by the **greenhouse effect** and **global warming**. Discuss the causes and effects (on ecosystems) of increased greenhouse gas emissions.

☐ 9. Recognise the importance of **conserving resources**. Using examples, describe the effects on ecosystems of resource harvesting. Discuss how humans can reduce their impact on ecosystems by managing resources sustainably, both on a local and international level.

☐ 10. Explain what is meant by the term **biodiversity** and discuss the importance of preserving and managing it as you would any resource. Identify and discuss factors important in the decline of the world's biodiversity. Identify measures to halt this decline.

☐ 11. Distinguish between **threatened** and **endangered species** and identify factors that cause species to become endangered. Identify some endangered species and predict the short and long term consequences of species loss.

See page 7 for additional details of these texts:

■ Allen, D, M. Jones, and G. Williams, 2001. **Applied Ecology**, pp. 18-19, 26-33, 90-98

■ Reiss, M. & J. Chapman, 2000. **Environmental Biology** (Cambridge University Press), pp. 5-7, 9-12, 24-25, chpt. 3.

■ Smith, R. L. & T.M. Smith, 2001. **Ecology and Field Biology**, reading as required.

See page 7 for details of publishers of periodicals:

■ **The Impact of Habitat Fragmentation on Arthropod Biodiversity** The American Biology Teacher, 62(6), June 2000, pp. 414-420. *An account of experimental work to investigate the impact of human activity on arthropod populations.*

■ **Biodiversity and Ecosystems** Biol. Sci. Rev., 11(4) March 1999, pp. 18-21. *The importance of biodiversity to ecosystem stability and sustainability.*

■ **Hot Spots** New Scientist, 4 April 1998, pp. 32-36. *An examination of the reasons for the very high biodiversity observed in the tropics.*

■ **Unlocking the Climate Puzzle** National Geographic, 193(5) May 1998, pp. 38-71. *Earth's climate, including global warming & desertification.*

■ **Biodiversity: Taking Stock of Life** National Geographic, 195(2) February 1999 (entire issue). *Special issue exploring the Earth's biodiversity and what we can do to preserve it.*

■ **Down on the Farm: The Decline in Farmland Birds** Biol. Sci. Rev., 16(4) April 2004, pp. 17-20. *Factors in the decline of bird populations in the UK.*

■ **How Did Humans First Alter Global Climate?** Sci. American, March 2005, pp. 34-41. *A new hypothesis suggests that humans began altering the global climate thousands of years before their more recent use of fossil fuels.*

■ **The Big Thaw** National Geographic, Sept. 2004, pp. 12-75. *Part of a special issue providing an up-to-date, readable account of the state of global warming and climate change.*

■ **The Case of the Missing Anurans** The American Biology Teacher, 63(9), Nov. 2001, pp. 670-676. *The threat to the world's frog populations; an article investigating threatened species decline.*

■ **Plant Succession** Biol. Sci. Rev., 14 (2) Nov. 2001, pp. 2-6. *Primary and secondary succession, including causes of different types of succession.*

See pages 4-5 for details of how to access **Bio Links** from our web site: **www.thebiozone.com**. From Bio Links, access sites under the topics:

BIODIVERSITY > Biodiversity: • Biolink Australia: Biodiversity ... *and others* **CONSERVATION:** > **Habitat Loss:** • Rainforest Information Centre > **Conservation Issues:** • Environment Australia online ... *and others* **HUMAN IMPACT:** > **Pollution:** • Urban and regional air pollution (Aus) ... *and others* > **Global Warming:** • The greenhouse effect (CSIRO) ... *and others*

Presentation MEDIA to support this topic:

ECOLOGY
• Biodiversity & Conservation
• Human Impact

Ecology

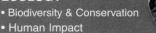

Environmental Change

Natural environmental changes come from three sources: the **biosphere** itself, **geological forces** (crustal movements and plate tectonics), and **cosmic forces** (the movement of the moon around the Earth, and the Earth and planets around the sun). All three forces can cause cycles, steady states, and trends (directional changes) in the environment. Environmental trends (such as climate cooling) cause long term changes in communities. Some short term cycles may also influence patterns of behaviour and growth in many species. These cyclical behaviour patterns, or **biological rhythms**, are based around regular environmental changes, which are used as **cues** to regulate the timing of the behaviour.

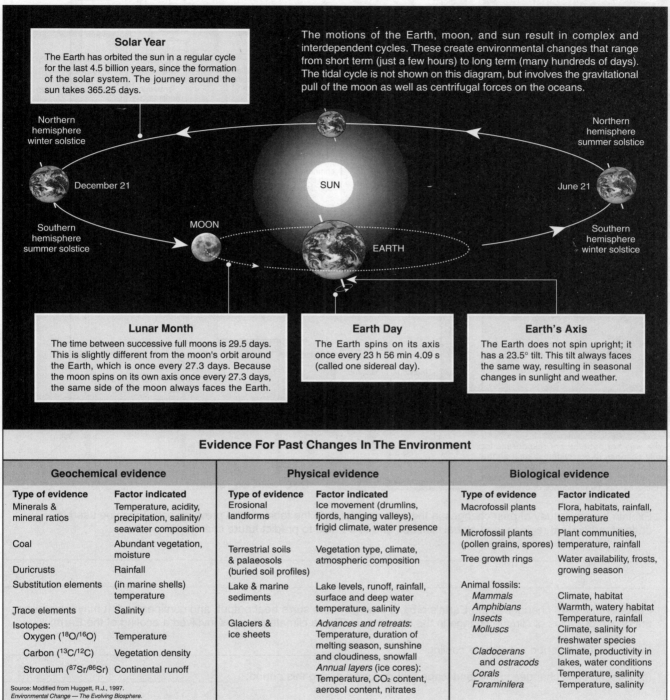

Solar Year

The Earth has orbited the sun in a regular cycle for the last 4.5 billion years, since the formation of the solar system. The journey around the sun takes 365.25 days.

The motions of the Earth, moon, and sun result in complex and interdependent cycles. These create environmental changes that range from short term (just a few hours) to long term (many hundreds of days). The tidal cycle is not shown on this diagram, but involves the gravitational pull of the moon as well as centrifugal forces on the oceans.

Northern hemisphere winter solstice

Northern hemisphere summer solstice

December 21

SUN

June 21

Southern hemisphere summer solstice

MOON

EARTH

Southern hemisphere winter solstice

Lunar Month

The time between successive full moons is 29.5 days. This is slightly different from the moon's orbit around the Earth, which is once every 27.3 days. Because the moon spins on its own axis once every 27.3 days, the same side of the moon always faces the Earth.

Earth Day

The Earth spins on its axis once every 23 h 56 min 4.09 s (called one sidereal day).

Earth's Axis

The Earth does not spin upright; it has a 23.5° tilt. This tilt always faces the same way, resulting in seasonal changes in sunlight and weather.

Evidence For Past Changes In The Environment

Geochemical evidence		Physical evidence		Biological evidence	
Type of evidence	**Factor indicated**	**Type of evidence**	**Factor indicated**	**Type of evidence**	**Factor indicated**
Minerals & mineral ratios	Temperature, acidity, precipitation, salinity/ seawater composition	Erosional landforms	Ice movement (drumlins, fjords, hanging valleys), frigid climate, water presence	Macrofossil plants	Flora, habitats, rainfall, temperature
Coal	Abundant vegetation, moisture	Terrestrial soils & palaeosols (buried soil profiles)	Vegetation type, climate, atmospheric composition	Microfossil plants (pollen grains, spores)	Plant communities, temperature, rainfall
Duricrusts	Rainfall			Tree growth rings	Water availability, frosts, growing season
Substitution elements	(in marine shells) temperature	Lake & marine sediments	Lake levels, runoff, rainfall, surface and deep water temperature, salinity		
Trace elements	Salinity			Animal fossils:	
Isotopes:		Glaciers & ice sheets	*Advances and retreats*: Temperature, duration of melting season, sunshine and cloudiness, snowfall	Mammals	Climate, habitat
Oxygen ($^{18}O/^{16}O$)	Temperature			Amphibians	Warmth, watery habitat
Carbon ($^{13}C/^{12}C$)	Vegetation density			Insects	Temperature, rainfall
Strontium ($^{87}Sr/^{86}Sr$)	Continental runoff		*Annual layers* (ice cores): Temperature, CO_2 content, aerosol content, nitrates	Molluscs	Climate, salinity for freshwater species
				Cladocerans and *ostracods*	Climate, productivity in lakes, water conditions
				Corals	Temperature, salinity
				Foraminifera	Temperature, salinity

Source: Modified from Huggett, R.J., 1997.
Environmental Change — The Evolving Biosphere.

1. Describe one biological role of **environmental cues**: _____

   _____

2. For each of the following categories, provide an example of a type of evidence for environmental change:

   (a) Geochemical: _____

   (b) Physical: _____

   (c) Biological: _____

Changes in Ecosystems

Code: A 2

Time scale and geographic extent of environmental change
(Time scale: horizontal axis / geographic extent: vertical axis)

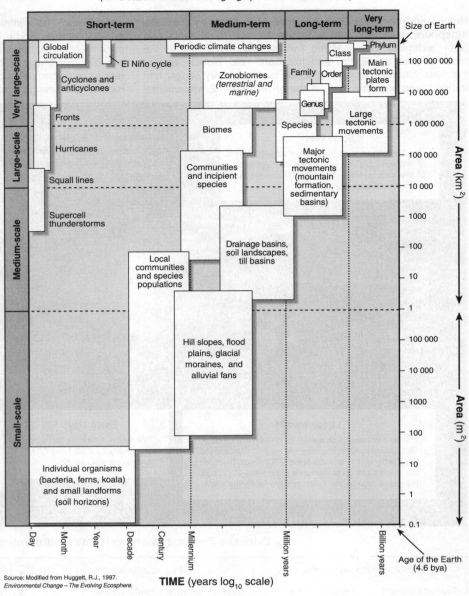

Climatic change during the last 2-3 million years has involved cycles of glacial and interglacial conditions. These cycles are largely the result of an interplay between astronomical cycles and atmospheric CO_2 concentrations.

Volcanic eruptions may have a large effect on local biological communities. They may also cause prolonged changes to regional and global weather (e.g. Mount Pinatubo eruption, 1989).

Some weather patterns are responsible for subtle changes to ecosystems, such as the gradual onset of a drought. They may also provide large scale and forceful changes, such as those caused by hurricanes or cyclones.

Source: Modified from Huggett, R.J., 1997.
Environmental Change – The Evolving Ecosphere.

TIME (years log_{10} scale)

3. Explain how the study of past changes in the environment (see the table on the previous page) may be used to understand current regional and global climates and to be able to predict future changes:

_____

_____

4. Periodic long term changes in the Earth's orbit, a change in the sun's heat output, and continental drift may have been the cause of cycles of climate change in the distant past. These climate changes involved a cooling of the Earth:

(a) Name these periods of climate cooling: _____

(b) Describe two changes to the landscape that occurred during this period: _____

_____

_____

5. On the diagram above, colour-code each of the rectangles to indicate the four themes of environmental change:

Climatic: [ ]   Ecological: [ ]   Tectonic: [ ]   Evolutionary: [ ]

6. Explain to what extent these four types of environmental change are interlinked: _____

_____

_____

Ecosystem Stability

Ecological theory suggests that all species in an ecosystem contribute in some way to ecosystem function. Therefore, species loss past a certain point is likely to have a detrimental effect on the functioning of the ecosystem and on its ability to resist change (its **stability**). Although many species still await discovery, we do know that the rate of species extinction is increasing. Scientists estimate that human destruction of natural habitat is driving up to 100 000 species to extinction every year. This substantial loss of biodiversity has serious implications for the long term stability of many ecosystems.

The Concept of Ecosystem Stability

The stability of an ecosystem refers to its apparently unchanging nature over time. Ecosystem stability has various components, including **inertia** (the ability to resist disturbance) and **resilience** (ability to recover from external disturbances). Ecosystem stability is closely linked to the biodiversity of the system, although it is difficult to predict which factors will stress an ecosystem beyond its range of tolerance. It was once thought that the most stable ecosystems were those with the greatest number of species, since these systems had the greatest number of biotic interactions operating to buffer them against change. This assumption is supported by experimental evidence but there is uncertainty over what level of biodiversity provides an insurance against catastrophe.

Monoculture

Natural grassland

Rainforest

Deforestation

Single species crops (monocultures), such as the soy bean crop (above, left), represent low diversity systems that can be vulnerable to disease, pests, and disturbance. In contrast, natural grasslands (above, right) may appear homogeneous, but contain many species which vary in their predominance seasonally. Although they may be easily disturbed (e.g. by burning) they are very resilient and usually recover quickly.

Tropical rainforests (above, left) represent the highest diversity systems on Earth. Whilst these ecosystems are generally resistant to disturbance, once degraded, (above, right) they have little ability to recover. The biodiversity of ecosystems at low latitudes is generally higher than that at high latitudes, where climates are harsher, niches are broader, and systems may be dependent on a small number of key species.

Community Response to Environmental Change

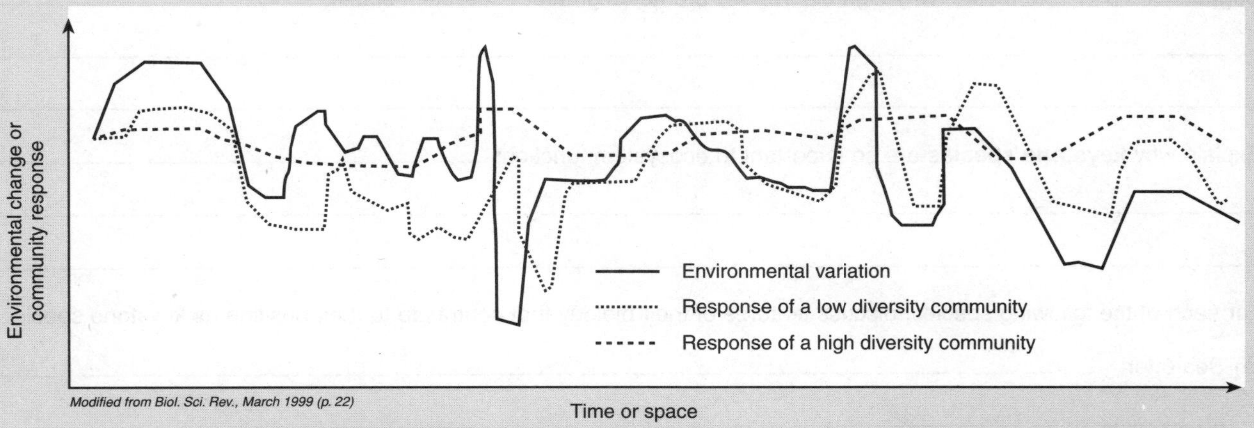

Graph: x-axis labelled "Time or space", y-axis labelled "Environmental change or community response"

Legend:
— Environmental variation
··· Response of a low diversity community
- - - Response of a high diversity community

Modified from Biol. Sci. Rev., March 1999 (p. 22)

In models of ecosystem function, higher species diversity increases the stability of ecosystem functions such as productivity and nutrient cycling. In the graph above, note how the low diversity system varies more consistently with the environmental variation, whereas the high diversity system is buffered against major fluctuations. In any one ecosystem, some species may be more influential than others in the stability of the system. Such **keystone (key) species** have a disproportionate effect on ecosystem function due to their pivotal role in some ecosystem function such as nutrient recycling or production of plant biomass.

Elephants can change the entire vegetation structure of areas into which they migrate. Their pattern of grazing on taller plant species promotes a predominance of lower growing grasses with small leaves.

Termites are amongst the few larger soil organisms able to break down plant cellulose. They shift large quantities of soil and plant matter and have a profound effect on the rates of nutrient processing in tropical environments.

The starfish *Pisaster* is found along the coasts of North America where it feeds on mussels. If it is removed, the mussels dominate, crowding out most algae and leading to a decrease in the number of herbivore species.

Keystone Species in North America

Gray wolf

Beaver, *Castor canadensis*

Sea otter, *Enhydra lutris*

Quaking aspen

Gray or timber wolves (*Canis lupus*) are a keystone predator and were once widespread in North American ecosystems. Historically, wolves were eliminated from Yellowstone National Park because of their perceived threat to humans and livestock. As a result, elk populations increased to the point that they adversely affected other flora and fauna. Wolves have since been reintroduced to the park and balance is returning to the ecosystem.

Two smaller mammals are also important keystone species in North America. Beavers (top) play a crucial role in biodiversity and many species, including 43% of North America's endangered species, depend partly or entirely on beaver ponds. Sea otters are also critical to ecosystem function. When their numbers were decimated by the fur trade, sea urchin populations exploded and the kelp forests, on which many species depend, were destroyed.

Quaking aspen (*Populus tremuloides*) is one of the most widely distributed tree species in North America, and aspen communities are among the most biologically diverse in the region, with a rich understorey flora supporting an abundance of wildlife. Moose, elk, deer, black bear, and snowshoe hare browse its bark, and aspen groves support up to 34 species of birds, including ruffed grouse, which depends heavily on aspen for its winter survival.

1. Suggest one probable reason why high biodiversity promotes greater ecosystem stability: _____

_____

_____

2. Explain why **keystone species** are so important to ecosystem function: _____

_____

_____

3. For each of the following species, discuss features of their biology that contribute to their position as keystone species:

(a) Sea otter: _____

_____

(b) Beaver: _____

_____

(c) Gray wolf: _____

_____

(d) Quaking aspen: _____

4. Giving examples, explain how the actions of humans to remove a keystone species might result in ecosystem change:

_____

_____

_____

_____

Ecological Succession

Ecological succession is the process by which communities in a particular area change over time. Succession takes place as a result of complex interactions of biotic and abiotic factors. Early communities modify the physical environment. This change results in a change in the biotic community which further alters the physical environment and so on. Each successive community makes the environment more favourable for the establishment of new species. A succession (called a **sere**) proceeds in stages, called seral stages, until the formation of the final climax community, which is stable until further disturbance. Early successional communities are characterised by having a low species diversity, a simple structure, and broad niches. In contrast, community structure in climax communities is complex, with a large number of species interactions. Niches are usually narrow and species diversity is high.

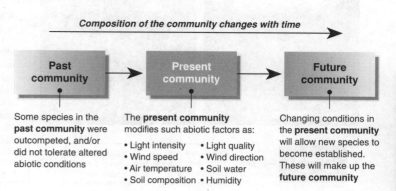

Composition of the community changes with time

Past community	Present community	Future community

Some species in the **past community** were outcompeted, and/or did not tolerate altered abiotic conditions

The **present community** modifies such abiotic factors as:
- Light intensity
- Light quality
- Wind speed
- Wind direction
- Air temperature
- Soil water
- Soil composition
- Humidity

Changing conditions in the **present community** will allow new species to become established. These will make up the **future community**

Changes in Ecosystems

Primary Succession

Primary succession refers to colonisation of regions where there is no preexisting community. Examples include the emergence of new volcanic islands, new coral atolls, or islands where the previous community has been extinguished by a volcanic eruption (e.g. the Indonesian island of Krakatau). A classic set colonisation sequence (as depicted below) is relatively uncommon. In reality, the rate at which plants colonise bare ground and the sequence of plant communities that develop are influenced by the local conditions and the dispersal mechanisms of plants in the surrounding region.

Climax community

Bare rock or volcanic ash	→	Lichens	→	Mosses and liverworts	→	Ferns and grasses	→	Shrubs, includng nitrogen fixers	→	Mature trees

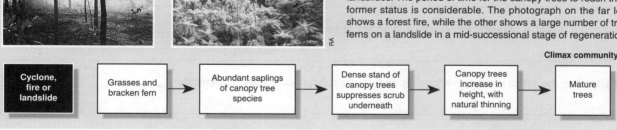

Secondary Succession: Catastrophe Cycle
(800 years for full richness to develop again)

Events that can cause localised or widespread destruction of forest include cyclone damage, forest fires, and hillside slips. The degree of destruction may vary from the loss of canopy trees in storms, to severe species depletion in the case of fire or landslides. The period of time for the canopy trees to reach their former status is considerable. The photograph on the far left shows a forest fire, while the other shows a large number of tree ferns on a landslide in a mid-successional stage of regeneration.

Climax community

Cyclone, fire or landslide	→	Grasses and bracken fern	→	Abundant saplings of canopy tree species	→	Dense stand of canopy trees suppresses scrub underneath	→	Canopy trees increase in height, with natural thinning	→	Mature trees

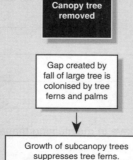

Secondary Succession: Gap Regeneration
(up to 500 years)

Large canopy trees have a profound effect on the make-up of the forest community immediately below. The reduced sunlight impedes the growth of most saplings. When a large tree falls, it opens a crucial hole in the canopy that lets in sunlight. There begins a race between the saplings to grow fast enough to fill the gap. The photograph on the left shows a large canopy tree in temperate rainforest that has recently fallen, leaving a gap in the canopy through which light can penetrate.

Canopy tree removed

Gap created by fall of large tree is colonised by tree ferns and palms

Climax community

Growth of subcanopy trees suppresses tree ferns. Seedlings of canopy trees grow beneath the subcanopy.	→	Rapid growth of saplings to occupy the gap	→	Mature trees develop to form climax community

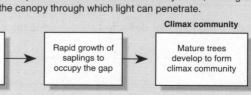

Secondary Succession in Cleared Land
(150+ years for mature woodland to develop again)

A secondary succession takes place after a land clearance (e.g. from fire or landslide). Such events do not involve loss of the soil and so tend to be more rapid than primary succession, although the time scale depends on the species involved and on climatic and edaphic (soil) factors. Humans may deflect the natural course of succession (e.g. by mowing) and the climax community that results will differ from the natural community. A climax community arising from a **deflected succession** is called a **plagioclimax**.

Pioneer community

Mature woodland

Climax community

Primarily bare earth	Open pioneer community (annual grasses)	Grasses and low growing perennials	Scrub: shrubs and small trees	Young broad-leaved woodland	Mature woodland mainly oak

Time to develop (years)   1–2   3–5   16-30   31-150   150 +

1. Distinguish between **primary** succession and **secondary** succession: _____

_____

_____

_____

_____

2. Suggest why primary successions rarely follow the classic sequence depicted on the previous page: _____

_____

_____

_____

_____

3. (a) Identify some early colonisers during the establishment phase of a community on bare rock: _____

_____

    (b) Describe two important roles of the species that are early colonisers of bare slopes: _____

_____

_____

_____

_____

4. Describe a possible catastrophic event causing succession in a rainforest ecosystem: _____

_____

5. (a) Describe the effect of selective logging on the composition of a forest community: _____

_____

_____

    (b) Suggest why selective logging could be considered preferable (for forest conservation) to clear felling of trees:

_____

_____

_____

6. (a) Explain what is meant by a **deflected succession**: _____

_____

    (b) Discuss the role that deflected successions might have in maintaining managed habitats: _____

_____

_____

_____

_____

Wetland Succession

Wetland areas present a special case of ecological succession. Wetlands are constantly changing as plant invasion of open water leads to siltation and infilling. This process is accelerated by **eutrophication**. In well drained areas, pasture or **heath** may develop as a result of succession from freshwater to dry land. When the soil conditions remain non-acid and poorly drained, a swamp will eventually develop into a seasonally dry **fen**. In special circumstances (see below) an acid **peat bog** may develop. The domes of peat that develop produce a hummocky landscape with a unique biota. Wetland peat ecosystems may take more than 5000 years to form but are easily destroyed by excavation and lowering of the water table.

Wetland Succession

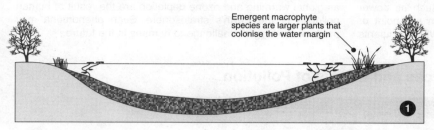

Emergent macrophyte species are larger plants that colonise the water margin

An open body of water, with time, becomes silted up and is invaded by aquatic plants. Emergent macrophyte species colonise the accumulating sediments, driving floating plants towards the remaining deeper water.

Deep water submergent plant species

The increasing density of rooted emergent, submerged, and floating macrophytes encourages further sedimentation by slowing water flows and adding organic matter to the accumulating silt.

Encroaching margins

The resulting **swamp** is characterised by dense growths of emergent macrophytes and permanent (although not necessarily deep) standing water. As sediment continues to accumulate the swamp surface may, in some cases, dry off in summer.

Peat dome

In colder climates, low evaporation rates, high rainfall, and invasion by *Sphagnum* moss leads to development of a peat **bog**; a low pH, nutrient poor environment where acid-tolerant plants such as sundew replace the less acid-tolerant swamp species.

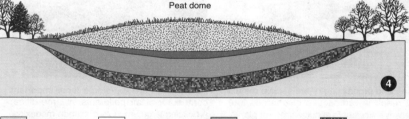

| | Bedrock | | Open water | | Lake muds | | Lake clay | | Swamp peat | | Bog peat |

1. Identify the two most important abiotic factors in determining whether or not a wetland will develop into bog:

   (a) _____ (b) _____

2. Describe where these conditions encouraging bog development are found: _____

   _____

3. In many areas, bogs are threatened by drainage of land for pasture. With respect to the abiotic environment, suggest why land drainage threatens the viability of bog ecosystems:

   _____

   _____

   _____

4. Describe how *Sphagnum* alters the environment to favour the establishment of other bog species: _____

   _____

   _____

Code: A 2

Pollution

Any addition to the air, water, soil, or food that threatens the survival, health, or activities of organisms is called **pollution**. **Pollutants** can enter the environment naturally (e.g. from volcanic eruptions) or through human activities. Most pollution from human activity occurs in or around urban and industrial areas and regions of industrialised agriculture. Pollutants may come from single identifiable **point sources**, such as power plants, or they may enter the environment from non-point or **diffuse sources**, such as through land runoff. While pollutants often contaminate the areas where they are produced, they can also be carried by wind or water to other areas. Commonly recognised forms of pollution include air pollution, water pollution, and soil contamination, but other less obvious forms of pollution, including light and noise pollution, are also the result of concentrations of human activity. Some global phenomena, such as global warming and ozone depletion are the result of human pollution of the Earth's stratosphere. Such phenomena may produce the biggest challenge to humans in the future.

Sources and Effects of Pollution

Soil contamination occurs via chemical spills, leaching, or leakage from underground storage. The runoff from open-caste mining operations can be loaded with heavy metals such as mercury, cadmium, and arsenic.

Sewage: Water containing human wastes, soaps and detergents, pathogens, and toxins are discharged into waterways and the sea. Most communities apply some level of waste water treatment at sewage treatment facilities.

Power plants and industrial emissions are a major source of air pollution. SO_2 and NO_2 from these primary sources mix with water vapour in the atmosphere to form acids which may be deposited as rain, snow, or dry acid.

Fertilisers, herbicides, and pesticides are major contaminants of soil and water in areas where agriculture is industrialised. Fertiliser runoff and leaching adds large quantities of nitrogen and phosphorus to waterways and leads to accelerated **eutrophication**.

Mining and processing of radioactive metals may result in radioactive discharges into waterways. Accidents at nuclear power plants, such as at Chornobyl in 1986, (above) can cause widespread radioactive contamination which persists for long periods of time.

Automobiles are the single most important contributor of air pollutants in large cities, producing large amounts of carbon monoxide, hydrocarbons, and nitrous oxides. Ozone and smog are created as nitrogen oxides and hydrocarbons react to sunlight.

1. For each of the following forms of pollution, identify the source of the pollution and summarise its effects:

   (a) Accelerated nutrient enrichment and eutrophication of waterways: _____

   _____

   (b) Acid deposition: _____

   _____

   (c) Smog: _____

   _____

2. Explain why pollution from non-point sources is more difficult to identify and control than pollution form point sources:

   _____

   _____

3. Identify one important source of pollution in your region and, on a separate sheet, discuss its causes and effects, as well as any measures to control the pollution or mitigate against environmental damage. Attach your discussion to this page.

Global Warming

The Earth's atmosphere comprises a mixture of gases including nitrogen, oxygen, and water vapour. Also present are small quantities of carbon dioxide (CO_2), methane, and a number of other "trace" gases. In the past, our climate has shifted between periods of stable warm conditions to cycles of glacials and interglacials. The current period of warming climate is partly explained by the recovery after the most recent ice age that finished 10 000 years ago. Eight of the ten warmest years on record (records kept since the mid-1800s) were in the 1980s and 1990s. Global surface temperatures in 1998 set a new record by a wide margin, exceeding those of the previous record year, 1995. Many researchers believe the current warming trend has

been compounded by human activity, in particular, the release of certain gases into the atmosphere. The term '**greenhouse effect**' describes a process of global climate warming caused by the release of 'greenhouse gases', which act as a thermal blanket in the atmosphere, letting in sunlight, but trapping the heat that would normally radiate back into space. About three-quarters of the natural greenhouse effect is due to water vapour. The next most significant agent is CO_2. Since the industrial revolution and expansion of agriculture about 200 years ago, additional CO_2 has been pumped into the atmosphere. The effect of global warming on agriculture, other human activities, and the biosphere in general, is likely to be considerable.

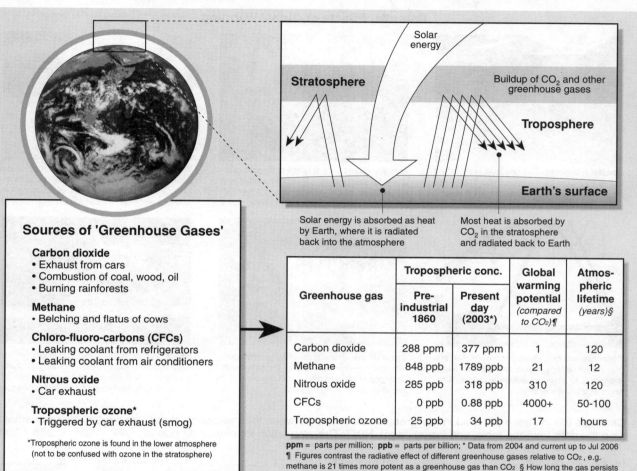

Sources of 'Greenhouse Gases'

Carbon dioxide
- Exhaust from cars
- Combustion of coal, wood, oil
- Burning rainforests

Methane
- Belching and flatus of cows

Chloro-fluoro-carbons (CFCs)
- Leaking coolant from refrigerators
- Leaking coolant from air conditioners

Nitrous oxide
- Car exhaust

Tropospheric ozone*
- Triggered by car exhaust (smog)

*Tropospheric ozone is found in the lower atmosphere (not to be confused with ozone in the stratosphere)

Greenhouse gas	Tropospheric conc.		Global warming potential (compared to CO_2)¶	Atmospheric lifetime (years)§
	Pre-industrial 1860	Present day (2003*)		
Carbon dioxide	288 ppm	377 ppm	1	120
Methane	848 ppb	1789 ppb	21	12
Nitrous oxide	285 ppb	318 ppb	310	120
CFCs	0 ppb	0.88 ppb	4000+	50-100
Tropospheric ozone	25 ppb	34 ppb	17	hours

ppm = parts per million; **ppb** = parts per billion; * Data from 2004 and current up to Jul 2006 ¶ Figures contrast the radiative effect of different greenhouse gases relative to CO_2, e.g. methane is 21 times more potent as a greenhouse gas than CO_2 § How long the gas persists in the atmosphere *Source: Carbon Dioxide Information Analysis Centre, Oak Ridge National Laboratory, USA.*

The graph on the right shows how the mean temperature for each year from 1860 until 2003 (grey bars) compared with the average temperature between 1961 and 1990. The thick black line represents the mathematically fitted curve and shows the general trend indicated by the annual data. Most anomalies since 1977 have been above normal; warmer than the long term mean, indicating that global temperatures are tracking upwards. In 1998 the global temperature exceeded that of the previous record year, 1995, by about 0.2°C.

Source: Hadley Centre for Prediction and Research

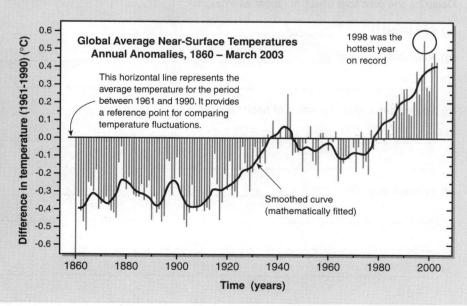

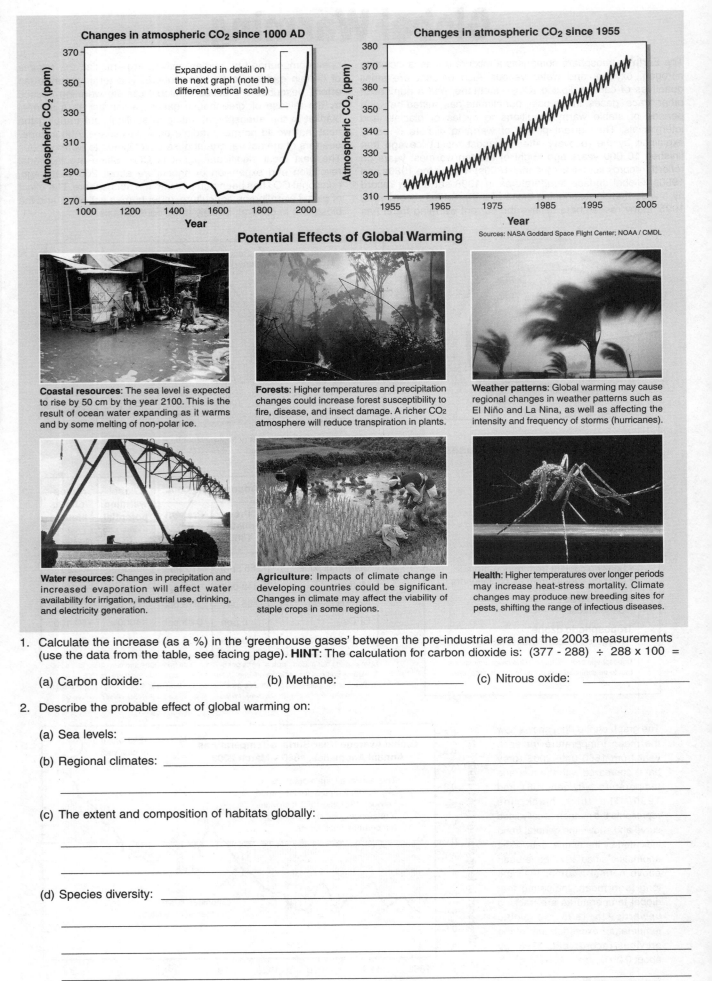

Changes in atmospheric CO₂ since 1000 AD

Atmospheric CO₂ (ppm)

Expanded in detail on the next graph (note the different vertical scale)

Year

Changes in atmospheric CO₂ since 1955

Atmospheric CO₂ (ppm)

Year

Sources: NASA Goddard Space Flight Center; NOAA / CMDL

Potential Effects of Global Warming

Coastal resources: The sea level is expected to rise by 50 cm by the year 2100. This is the result of ocean water expanding as it warms and by some melting of non-polar ice.

Forests: Higher temperatures and precipitation changes could increase forest susceptibility to fire, disease, and insect damage. A richer CO₂ atmosphere will reduce transpiration in plants.

Weather patterns: Global warming may cause regional changes in weather patterns such as El Niño and La Nina, as well as affecting the intensity and frequency of storms (hurricanes).

Water resources: Changes in precipitation and increased evaporation will affect water availability for irrigation, industrial use, drinking, and electricity generation.

Agriculture: Impacts of climate change in developing countries could be significant. Changes in climate may affect the viability of staple crops in some regions.

Health: Higher temperatures over longer periods may increase heat-stress mortality. Climate changes may produce new breeding sites for pests, shifting the range of infectious diseases.

1. Calculate the increase (as a %) in the 'greenhouse gases' between the pre-industrial era and the 2003 measurements (use the data from the table, see facing page). **HINT**: The calculation for carbon dioxide is: $(377 - 288) \div 288 \times 100 =$

   (a) Carbon dioxide: _____ (b) Methane: _____ (c) Nitrous oxide: _____

2. Describe the probable effect of global warming on:

   (a) Sea levels: _____

   (b) Regional climates: _____

   _____

   (c) The extent and composition of habitats globally: _____

   _____

   _____

   _____

   (d) Species diversity: _____

   _____

   _____

   _____

   _____

Loss of Biodiversity

The species is the basic unit by which we measure biological diversity or **biodiversity**. Biodiversity is not distributed evenly on Earth, being consistently richer in the tropics and concentrated more in some areas than in others. Conservation International recognises 25 **biodiversity hotspots**. These are biologically diverse and ecologically distinct regions under the greatest threat of destruction. They are identified on the basis of the number of species present, the amount of **endemism**, and the extent to which the species are threatened. More than a third of the planet's known terrestrial plant and animal species are found in these 25 regions, which cover only 1.4% of the Earth's land area. Unfortunately, biodiversity hotspots often occur near areas of dense human habitation and rapid human population growth. Most are located in the tropics and most are forests. Loss of biodiversity reduces the stability and resilience of natural ecosystems and decreases the ability of their communities to adapt to changing environmental conditions. With increasing pressure on natural areas from urbanisation, roading, and other human encroachment, maintaining species diversity is paramount and should concern us all today.

Biodiversity Hotspots

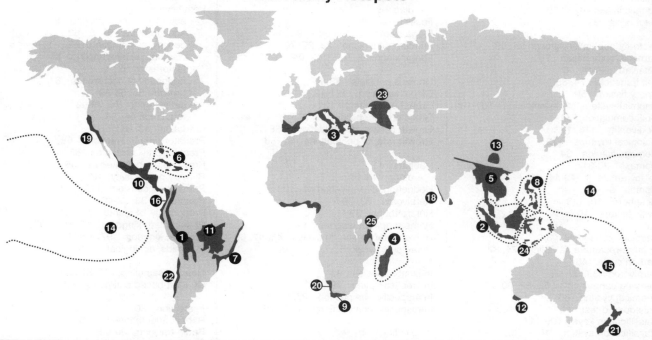

Threats to Biodiversity

Rainforests in some of the most species-rich regions of the world are being destroyed at an alarming rate as world demand for tropical hardwoods increases and land is cleared for the establishment of agriculture.

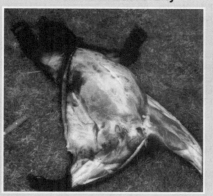

Illegal trade in species (for food, body parts, or for the exotic pet trade) is pushing some species to the brink of extinction. Despite international bans on trade, illegal trade in primates, parrots, reptiles, and big cats (among others) continues.

Pollution and the pressure of human populations on natural habitats threatens biodiversity in many regions. Environmental pollutants may accumulate through food chains or cause harm directly, as with this bird trapped in oil.

1. Use your research tools (e.g. textbook, internet, or encyclopaedia) to identify each of the 25 biodiversity hotspots illustrated in the diagram above. For each region, summarise the characteristics that have resulted in it being identified as a biodiversity hotspot. Present your summary as a short report and attach it to this page of your workbook.

2. Identify the threat to biodiversity that you perceive to be the most important and explain your choice:

_____

_____

_____

Changes in Ecosystems

Code: RA 3

Index